Praise for *In Search Of Real Monsters*

"Richard Freeman must surely be the world's most widely-travelled field cryptozoologist, scouring the globe for well over twenty years in search of such elusive mystery beasts as British lake monsters, Mongolian death worms, Tasmanian wolves, Russian and Sumatran man-beasts, South American water tigers, and African dragons, to name but a few. Now he has drawn upon the extensive knowledge and experience gained during his many expeditions and voracious reading to write this fascinating book, packed with new, original insights and forthright opinions, making it essential reading for everyone who dreams of following in his footsteps seeking unknown animals."

—Dr. Karl Shuker, zoologist and author

"Freeman's book on monsters is a page-turning treat. Not only does he pack it with excellent overviews, facts, and details on cryptozoology, he also grips the reader with accounts of his own expeditions. These globe-trotting journals bring the hidden and mysterious creatures of nature into the mud and dirt of serious, real-world research, leaving us with a thrilling and inspiring book. Yes, Freeman takes monsters seriously, but he never loses his sense of fun, wonder, and adventure. Recommended."

—Peter Laws, author of *The Frighteners: Why We Love Monsters, Ghosts, Death, and Gore*

In Search
of
Real Monsters

Published by Mango Publishing Group, a division of Mango Media Inc.

Cover Design & Art Direction: Morgane Leoni
Cover Illustration: ruskpp / stock.adobe.com

For permission requests, please contact the publisher at:
Mango Publishing Group
2850 S Douglas Road, 4th Floor
Coral Gables, FL 33134 USA
info@mango.bz

For special orders, quantity sales, course adoptions and corporate sales, please email the publisher at sales@mango.bz. For trade and wholesale sales, please contact Ingram Publisher Services at customer.service@ingramcontent.com or +1.800.509.4887.

In Search of Real Monsters: Adventures in Cryptozoology Volume 2

Library of Congress Cataloging-in-Publication number: 2021946732
ISBN: (p) 978-1-64250-750-8 (e) 978-1-64250-751-5
BISAC NAT007000, NATURE / Animals / Dinosaurs & Prehistoric Creatures

Printed in the United States of America

In Search
of
Real Monsters

Adventures in Cryptozoology

VOLUME II

Richard Freeman

CORAL GABLES

Table of Contents

Foreword

I started the Centre for Fortean Zoology back in 1992, but it was four years later that, at the Unconvention run by *Fortean Times*, I met Richard Freeman, who was then a student in zoology at Leeds University. And we have been friends ever since. He spent much of 1997 visiting me and my friend Graham Inglis at my house in Exeter. Together, we carried out a string of investigations into local zoological mysteries; we hunted for big cats, visited the woodlands in southern Cornwall which are the haunts of the grotesque owlman, and carried out a necropsy on a dolphin which had died under mysterious circumstances.

It seemed, therefore, totally logical that, when Richard finished his studies in 1998, he would move down to Exeter to work with us.

For the next seven years, until I moved up to North Devon to look after my dying father, Richard and I were housemates and planned his early expeditions. The first one was in 2000, when he went to Thailand, searching for the terrifying serpent dragon, the naga. Upon his return, I was on the edge of my seat as I heard how he crawled through subterranean passages in which he was probably the only European to have been, in the certain knowledge that if anything went wrong, nobody would be able to rescue him, and he would certainly die down there. Three years later came his first expedition in search of the Orang-Pendek, a fabled upright walking ape of the Sumatran rain forests. It was the first of many trips that he took to the Kerinci Sablat National Park, a series of expeditions which finally bore fruit in 2009, when Dave Archer and Sahar Deimus of the expedition actually saw the creature.

Since those early expeditions, Richard has been to the Gambia, Karbadino Balkaria in the Russian Caucasus, Northern India, Guyana, and on a number of trips to Tasmania in search of the thylacine.

Richard is more than a friend to me; he is closer to being a brother. And I have worked with him now for a quarter of a century. He is undoubtedly the most visible member of the Centre for Fortean Zoology, and as well as being an intrepid explorer who has visited many places where no other European has been, he is a fine author and raconteur, and—unusually, for the present generation of cryptozoologists—he is a very logical biological theorist. Not for him, ludicrous flights of fancy which would fit more obviously into works of science fiction, but firm and logical hypothesising which fit comfortably into the accepted framework of modern zoology.

People like Richard and myself find it hard to fit into twenty-first-century Britain. Richard in particular gives off the aura of somebody who was born a hundred years before, and in an increasingly homogenous and often bland society, in my humble opinion, we need people like Richard Freeman more than ever.

—Jon Downes
Director, Centre for Fortean Zoology
North Devon
July 2021

Introduction

So here we are at the start of volume two of *Adventures in Cryptozoology*. It was never meant to be a two-volume work, but the sheer amount of material, even in an introductory work such as this, would have resulted in a tome thick enough to use as a doorstop. Ergo the hefty volume was rent asunder into two parts.

In this part we continue our exploration of real-life monsters by first looking at creatures once known to have existed but now thought by mainstream science to have become extinct. In many cases, this presumption may be premature. Chief among these is the Tasmanian wolf, the beautiful animal that graces the cover of this very book—a wolf-like striped marsupial that haunts the wilds of Tasmania, and possibly mainland Australia and the mysterious island of New Guinea. In the jungles of South America, we follow the spoor of the giant ground sloth, a shaggy giant thought extinct since the end of the last ice age. Could relic populations still persist in the depths of the green hell of the Amazon and other remote areas?

Some monsters may have an even older pedigree. From the steaming rain forests and swamps of Central Africa come stories of surviving dinosaurs. Vast, long-necked sauropods, horned ceratopsians, and even the infamous *Tyrannosaurus rex*. Do they really exist, and if they do, are they really what we think they are?

Monsters do not need to be prehistoric or even unknown. Size itself can be a factor in creating a cryptid. We will examine hair-raising stories of giant reptiles. Crocodiles are feared as the most dangerous man-eaters on the planet, but we do not know the true extent of their size. From three continents, we look at stories of monster crocs that dwarf the largest known specimens and are truly the stuff of our most primal fears. We also go on the trail of mega-serpents, massive anacondas and pythons from the traveller's tales of old to modern-day sightings.

This volume is more personal than the first insofar as I recount some of my own adventures tracking cryptids across the globe. These include expeditions in search of the yeti, the Mongolian death worm, the Tasmanian wolf, the orang-pendek, the almasty, and the gul.

Finally, I will end the book with some advice for would-be cryptozoologists who want to follow in my footsteps. I will tell you how to organize your very own cryptozoological expedition from scratch. What beast to select? What equipment to take? How to find native guides? These and other questions will be answered.

Good luck—it may be you writing the next book on cryptozoology with your own discoveries.

—RICHARD FREEMAN,
THE CENTRE FOR FORTEAN ZOOLOGY,
DEVON, ENGLAND

"Extinction is the rule. Survival is
the exception."

—Carl Sagan

Time and again animals that we have thought to be extinct have turned up alive and well. The night parrot (*Pezoporus occidentalis*) is a small parrot from central Australia which was thought to be extinct. No sighting had occurred since 1912. Then, in 2013, it turned up out of the blue in a remote, arid area of the outback.

The night parrot is a small creature, a bit like a budgie on steroids. Not like the Zanzibar leopard (*Panthera pardus pardus*), which was thought extinct after a campaign to exterminate it in the 1960s. In 2018 one strolled into view and was filmed by a camera trap. Zanzibar has an area of 950 square miles, a tiny space in which to hide a population of leopards, but they are there.

The takahē (*Porphyrio hochstetteri*) is a large, flightless bird related to the moorhen and found only on New Zealand. It was last seen in 1898, but was rediscovered in 1948 in the mountains of South Island.

The Bermuda petrel (*Pterodroma cahow*) is a seabird identified in 1612. It was thought to have been hunted to extinction by the 1620s, and was thought extinct for over three hundred years, until one crashed into a lighthouse in 1951. We now know this "extinct" bird nests on four islands off Bermuda.

These are just a few examples from a long list. If these creatures could survive, we must ask ourselves, what other animals that mainstream science assures us are long gone still walk the earth?

The Tasmanian Wolf

Of all the world's cryptids, the most likely to exist is the enigmatic and beautiful creature known as the thylacine. This flesh-eating marsupial is one of the most spectacular examples of convergent evolution, where two distinct species, often on opposite sides of the world, bear a remarkable resemblance to one another due to both inhabiting similar ecological niches. The thylacine (*Thylacinus cynocephalus*) is also known as the Tasmanian wolf or Tasmanian tiger and is convergent with the placental wolf. The animal bears a striking resemblance to a wolf or dog, but with stripes along its hindquarters. Of course, it is not related to the wolf or the tiger. Neither should it be confused with the Tasmanian devil (*Sarcophilus harrisii*), a superficially badger-like flesh-eating marsupial, or the spotted or tiger quoll (*Dasyurus maculatus*), a native cat-like marsupial predator. Both sexes have a backward-facing pouch. In females it is used to nurture and protect developing young, and in males to protect the sex organs as the animal runs through vegetation after prey. The skull has a gape far wider than that of a wolf or dog. The thylacine's dental formula is different to a wolf's. It bears four incisors and four molars in each quadrant of the jaw, as opposed to only three of each in true canids. The thylacine has a more powerful bite than a wolf, but the skull is less adapted to holding struggling prey. This suggests a different hunting strategy. Whereas pack-hunting wolves use number to pull down prey and worry it to death, thylacines may kill small prey animals with one bite and, with larger victims, inflict a bite, then let them bleed to death. It is not as well adapted for fast running as a wolf, but seems to have more stamina for pursuit over long distances.

The thylacine is the largest marsupial predator of recent times, and has a lineage that reaches back to the Miocene epoch. Thylacines were once found across mainland Australia and New Guinea as well as Tasmania. Standard thinking would have us believe that the species died out on the mainland around three thousand years ago, perhaps from diseases transmitted by the introduced dingo. However, sightings persist in both Australia and New Guinea up to the present day.

When white settlers first colonized Tasmania in 1803, they began an act of ecological genocide. The largest broad-leafed trees on earth, the giant mountain ashes, were cut down. The Tasmanian black emus were hunted into extinction by the 1830s. The Tasmanian Aboriginal populations were decimated by hunting and disease. Their culture has almost entirely vanished, and only vestiges remain.

Areas of forest were cut down to allow the grazing of sheep. The Tasmanian wolf was an inconvenience for sheep farmers. Doubtless the creature did indeed kill some sheep. Slow-moving, placid targets are hard for predators to resist, but the claims of predation by some sheep farmers were on such a scale as to be physically impossible.

Like most politicians everywhere and at every time, the Tasmanian officials were self-serving cowards with knee-jerk reactions. To be seen as doing *something*, they offered bounties on thylacines from 1830 to 1909. The bounty was set at one dollar per head. During those years, 2,184 bounties were paid. By the 1920s, the thylacine had become scarce in the wild. A thylacine was shot by Wilf Batty at Mawbanna in 1930. Elias Churchill trapped one alive in the Florentine Valley in 1933.

Many specimens were caught for zoos around the world, including London, but no concerted attempt was made to captive-breed them. The last captive animal died on September 7, 1936, at Hobart Zoo, apparently from cold, as it had been locked out of its sleeping quarters.

Since the date of the Tasmanian wolf's official extinction, there have been more than four thousand reported sightings.

These come not just from laymen, but also from some very credible witnesses, including zoologist Hans Naarding, who in 1982 observed a large male thylacine near the Arthur River in the state's northwest. He had spent decades studying animals around the world. In Tasmania, he had been studying a bird called Latham's snipe (*Gallinago hardwickii*). At two in the morning, he awoke.

"I was in the habit of intermittently shining a spotlight around. The beam fell on an animal in front of the vehicle, less than ten meters away. Instead of risking movement by grabbing for a camera, I decided to register very carefully what I was seeing. The animal was about the size of a small shepherd dog, a very healthy male in prime condition. What set it apart from a dog, though, was a slightly sloping hindquarter, with a fairly thick tail being a straight continuation of the backline of the animal. It had twelve distinct stripes on its back, continuing onto its butt. I

knew perfectly well what I was seeing. As soon as I reached for the
camera, it disappeared into the tea-tree undergrowth and scrub."

The official government report into the sighting concluded that "it must be accepted that thylacines survive in a number of areas of Tasmania."

Another expert witness was Charlie Beasley, a ranger with the Department of Environment and Land Management. It occurred in January 1995. Beasley was bird-watching at dusk in a valley in the Pyangana region, inland from St. Helens, in the northeast of the island. He saw an animal sniffing around on a ledge, and observed via binoculars and described the beast.

"Dirty brown colour with black stripes down its ribcage and
about half the size of a full-grown Alsatian dog. It had a face
like a Staffordshire bull terrier, but more elongated. The animal
stretched, turned, and walked back into the dense scrub. The tail
was heavy and somewhat like that of a kangaroo, and was held
out in a gentle curve."

Beasley had the animal in view for two minutes.

The creature's continued survival has even been predicted by a computer programme. Professor Henry Nix of the Australian University's Centre for Resource and Environmental Studies developed a programme called BIOCLIM. A research tool, BIOCLIM matched what was known about the habits and preferences of a species with geographical areas. It matched the two up and predicted where, within a given area, the target species was most likely to be found. Nix applied this to the thylacine. There was an almost perfect match between where the programme predicted the animals would be if they had survived and the areas where sightings were being made. Nix concluded that people really were seeing thylacines. Professor Nix thought that as many as a thousand thylacines may still exist island-wide.

The following factors should also be noted. Firstly, there are many iconic extinct animals, such as the dodo (*Raphus cucullatus*), the great auk (*Pinguinus impennis*), and the passenger pigeon (*Ectopistes migratorius*), that nobody reports seeing. But people report the Tasmanian wolf on a regular basis. Secondly, the southwest of Tasmania was never settled, save for a handful of tin miners and fishermen at Port Davey. The area itself produced no thylacines during the bounty period. The area is not ideal for the animal, but we know that creatures under pressure can retreat to, and indeed thrive in, less-than-perfect conditions. A good example is the recently discovered population of Bengal tigers (*Panthera tigris tigris*) living in the high Himalayan mountains in Bhutan at an altitude of up to 11,500 feet, far above their normal range. Therefore, it is quite possible that thylacine populations moved into the southwest during the bounty years and remained unmolested. Eventually, these would have recolonized other areas of the island. Today most reports come from the northeast and west of Tasmania, and the west coast.

Dr. David Pemberton, curator of zoology at the Tasmanian Museum and Art Gallery, whose PhD thesis was on the thylacine, says that despite the scientific belief that five hundred animals are required to sustain a population, the Florida panther is down to a dozen or so animals and, while it does have some inbreeding problems, is still ticking along. He said, "I'd take a punt and say that, if we manage to find a thylacine in the scrub, it means that there are fifty-plus animals out there."

The thylacine's closest living relative, the Tasmanian devil has recently had problems with a disease sweeping through its populations. Devil facial tumour disease is a form of transmissible cancer passed on through bites. It has affected 65 percent of Tasmania and caused an 80 percent reduction in populations in the affected areas. However, genetic research into the devils has suggested that the species would only need a base population of around twenty-five individuals to repopulate. If the Tasmanian devil has the genetic capacity to do this, then perhaps the Tasmanian wolf does as well. It is not without reason that the thylacine has been called "the healthiest extinct animal you will ever see."

Further examples of thylacine sightings in Tasmania will be looked at in a later chapter. Now we will look at claims of thylacines seen on mainland Australia, where conventional wisdom tells us they died out around three thousand years ago.

Journalist Samela Harris of *Naracoorte News* began to collect and investigate stories of animals resembling Tasmanian wolves reported from South Australia. In 1967, a group of children on a school bus saw a strange, striped, dog-like animal between Naracoorte and Lucindale. The mother of one of the witnesses, Mrs. Dawn Anderson, also began to collect sightings. Between them, they amassed many eyewitness accounts. Mrs. Anderson produced a drawing based on the school children's descriptions. The sketch shows the distinctive long hind feet of a thylacine.

In mid-1967, Mrs. Anderson and her son observed a thylacine for fifteen minutes as it moved along a ditch in a swamp. In February of the following year, she and fifteen other people in three cars tried to corner a thylacine in a reed bed, unsuccessfully. In March of the same year, she saw one crossing a paddock.

Samela Harris interviewed a witness called Jack Victory, a Parks Commission employee who had seen one such creature along the Younghusband Peninsula.

> *"I was about four hundred yards away, looking at birds through a telescope. I just didn't know what he was... He was a large animal, a bit like a fox and a bit like a kangaroo. But he was neither. He started to run along, loping gait. He had a dog's head and a large, tapering, rather stiff-looking tail. His torso was striped in grey. The rest of the body was brown.*
>
> *"When we got to the spot where we had seen him, we found his paw prints in the clay. They were about the size of my fist, and looked quite similar when I stuck my fist into the clay beside his imprint. We estimated his weight to be between 120 and 150 pounds. The animal's appearance fits only that of the thylacine."*

Tourist officer John Pocock was rounding up emus in long grass on a private wildlife reserve just outside of Rendlesham when he saw an animal observing him. "It was a weird-looking thing, with canine features in the upper part of the body and marsupial features, like a kangaroo, at the rear. It was striped like a tiger."

A commonwealth film crew was in the area at the time, filming wildlife, but by the time he had located the crew and brought the cameraman to the area, the creature had gone.

A creature that was seen around the hamlet of Ozenkadnook in southern Victoria was given the tongue-twisting name of the Ozenkadnook tiger by the media. Farmer Cyril Tucker tracked one in 1962 and came within sixty feet of it. He said it was larger than an Alsatian dog, with a low-slung body, a long, thick tail, and a kangaroo-like head. It was grey with black stripes on the rump. It ran off with a strange loping gait, the hind legs moving together.

Another time he came upon the creature with his dogs. He set the dogs on it, and it leapt away, making three big bounds on its hind legs, a mode of movement thylacines were known to use. Tucker was lucky the beast did not turn on its pursuers. Thylacines have been known to bite right through the skulls of dogs that attack them.

In the same year, nine members of the Edenhope Hunt Club chased one of the animals through the scrub. Miss Lee Lightburn described it as "amazingly like a Tasmanian tiger."

In 1982, National Parks Ranger Peter Simon saw a thylacine in a clearing near Gibraltar Creek, Australian Capital Territory. Having seen many illustrations of the Tasmanian wolf, he was adamant that this is what he had seen. He was only one hundred feet from the animal as it crossed the clearing. During the following year, two groups of tourists told him that they had seen the same animal in the area.

Wilson's Promontory in Victoria is another hotspot for mainland sightings. It began in 1955, when something began to kill sheep in large numbers. Sheep were devoured overnight and dragged over two hundred yards. People started to report a strange creature that was named the Wonthaggi monster after a town in the era.

On December 6, 1955, Ern Featherstone, a car salesman, was demonstrating a car to Mr. and Mrs. T. J. Schmedje just one and a half miles from Wonthaggi when a strange creature appeared.

"It ran along the side of the road and disappeared into some scrub. When we stopped, it was looking at us. I've never seen anything like it. It was brown-striped, a sleek coat, and got along with a peculiar bound. It was two feet six inches tall and five feet long and had a tail as long as its body."

Mr. Schmedje added "It moved like a wallaby does when running on all fours. It had a fox-like head and long nose."

In November 1979, Mr. and Mrs. Charlie Thorpe were driving in the Promontory's National Park when a creature emerged from the bush and crossed the road in front of their car.

"We were not moving fast, probably around forty kilometres per hour, and got a good look at the animal. It was taller than my Labrador, but was lower in the hindquarters. It moved with a peculiar hopping gait. Its tail was very thick at the base and longer than a dog's, tapering to a point. It appeared to be a dark to light grey in colour and had distinctive darker bands around the hindquarters. The stripes did not appear to be black but were a darker grey than the rest of the body."

These are just a few scattered examples out of hundreds of sightings that suggest the creature may still be alive on the mainland. A number of films and photographs have turned up purporting to show thylacines on the mainland. Most of these appear to be feral red foxes (*Vulpes vulpes*) with mange.

In West Papua (formally Irian Jaya), the hill tribes report a dog-like carnivore they call dobsonga. They describe it as looking like a dog with striped flanks, a stiff tail, and wide jaws. They say it comes down from the mountains and kills pigs, goats, and other livestock. Thylacine hunter Ned Terry visited the area and showed the natives pictures of the Tasmanian wolf, which they identified as the dobsonga.

Ralf Kiesel, an explorer of Western Papua, wrote to renowned cryptozoologist Karl Shuker about persistent sightings of thylacines in Baliem Valley. In the early 1970s Jan Sarkang, a Papuan friend of Kiesel, working with a friend,

Punca Jaya, had just made camp for some geologists and were eating a meal. Two dog-like animals, an adult and a pup, emerged from the bush, apparently attracted by the smell of food. They were pale-coloured with wide mouths and stiff tails. The pup came close enough for one man to feed it. Then he tried to grab it, but the pup bit his hand and both animals ran back into the bush.

If I were a betting man, I'd put good money on the survival of the Tasmanian wolf. I think it is just a matter of time until definitive proof of the creature's continued existence comes to light. Maybe, by the time you are reading this book, Tasmania's most magnificent animal will have reemerged from the shadows into official existence once again. Interestingly, shortly after I returned from my first search for the thylacine, I came upon a very interesting book with a striking passage about the Tasmanian wolf. *The Sra Inside* was written by Philip Hoare, Visiting Fellow at the University of Southampton, and published by Fourth Estate in 2013. It consists of a number of essays on the world's oceans and on certain islands. In the chapter on Tasmania, the author writes extensively on the thylacine and modern-day sightings. He finishes the chapter with these words:

> *"What I do know is that in one institution I visit, a curator lets slip a quickly retracted remark, telling me it is not their secret to reveal. It is clear from what this person says, or does not say, that this strange half-life limbo of an animal which may or may not exist may soon be resolved, in its favour. That history is about to be reversed. That the thylacine is no longer extinct.*
>
> *"If it ever was."*

Living Dinosaurs

No group of animals that has ever lived holds us in such thrall as dinosaurs. They rampage through our childhood imaginings, and stalk across the silver screen from the jerky black-and-white silent *The Ghost of Slumber Mountain* (1918) to the high-tech *Jurassic Park*. There are dinosaur theme parks, dinosaur toys, dinosaur sweets, and dinosaur clothes, and more books have been written about dinosaurs than any other creatures.

The fascination is understandable. Humanity has been dominant on earth for less than one million years; dinosaurs were the unchallenged rulers of this sphere for more than one hundred and twenty million years! The cause of dinosaur extinction has still to be established. Palaeontologists are divided into two camps over this. The first favour the "smoking gun theories"—these evoke great global catastrophes to explain the terrible lizards' fall from supremacy. These include massive volcanic activity in Asia, radioactivity from exploding stars, and the ever-popular asteroid. Those in the second camp point out that the dinosaur decline was gradual and look to much more reasonable ideas, such as climate change and new diseases encountered whilst crossing newly formed land bridges as sea levels fell with global cooling.

The idea of total dinosaur extinction is totally false. Dinosaurs are around us all the time, and in some ways, they are just as successful now as they ever were—for birds are dinosaurs. As far back as 1860, when Archaeopteryx (then thought to be the first bird) was discovered in the Solenhofen shale of Bavaria, the link between dinosaurs and birds has been known. The specimen bore tailbones, teeth, and claws like any small carnivorous dinosaur, but beautifully preserved about it were unmistakable feathers. Since then, many feathered dinosaurs have been discovered—some pre-dating Archaeopteryx by several million years. All modern birds can literally be considered dinosaurs. Pneumatic bones, erect stance, and skull fenestrations are among their many shared features, and—strange as it may seem—*Tyrannosaurus rex*, the most powerful and savage predatory dinosaur, has more in common with your pet budgie than it does with Triceratops. When feeding ducks in the park, you are feeding dinosaurs. Your local pet shop sells dinosaurs for your home, and if you watch your uncle Ted eating chicken, you can boast to your friends you have seen a "man eating dinosaur." However, small feathery things that go "tweet" are not what the word "dinosaur" summons up for most people. Are there any non-avian dinosaurs surviving today? If so, they would make truly excellent dragons. Are there any giant sauropods or razor-toothed coelurosaurs slinking through the jungles or lurking on remote mesas awaiting formal discovery by incredulous scientists? There are those who think the answer is "yes!"

We begin our dinosaur safari in the cradle of mankind: darkest Africa.

Anyone boarding the Southampton train from Waterloo station on December 23, 1919, at 11:30 a.m., may well have been startled by two out-landish figures. One was a fierce-looking hound that seemed more wolf than dog. The other was a tall, weather-beaten man carrying a rifle. The man was Captain Leicester Stevens, and his dog was "Laddie"—a wolf-dog hybrid, a barrage-dog who had bravely carried messages under heavy fire in the First World War. His quest was to travel to Central Africa to hunt a surviving Brontosaurus. His intentions had made national news, and ironically—given what we now know about dinosaurs—an old lady from the Wild Birds Protection Association had written to him, asking him not to shoot the dinosaur. Sportsmen, hunters, and demobbed soldiers had written too, asking to accompany him. Perhaps unwisely, he elected to go alone—save for his dog.

The pair made it to the jungle, but were never seen again. No one knows what became of them. Without the adequate backup that a team expedition would have provided, they probably fell victim to tropical illness and died alone—thousands of miles from home. Sadly, it seems the report that had inspired their endeavour was a hoax. It appeared on November 17, 1919, in the *Times*.

A TALE FROM AFRICA

Semper aliqud novi

The Central News Port Elizabeth correspondent sends the following:

The head of the local museum here has received information from a M. Lepage, who was in charge of railway construction in the Belgian Congo, of an exciting adventure last month. While Lepage was hunting one day in October he came upon an extraordinary monster, which charged at him. Lepage fired but was forced to flee, with the monster in chase. The animal before long gave up the chase, and Lepage was able to examine it through his binoculars. The animal, he says, was about twenty-four feet in length, with a long pointed snout adorned with tusks like horns, and a short horn above the nostrils.

The front feet were like those of a horse, and the hind hooves were cloven. There was a scaly hump on the monster's shoulders.

The animal later charged through the native village of Fungurume, destroying the huts and killing some native dwellers. A hunt was organised but the government has forbidden the molestation of the animal, on the ground that it is probably a relic of antiquity. There is a wild trackless region in the neighbourhood which contains many swamps and marshes, where, says the head of the museum, it is possible a few primeval monsters may survive.

Firstly, the animal described does not resemble a Brontosaurus. This creature was a sauropod dinosaur, a long-necked herbivore. The creature Lepage reported more closely tallies with a ceratopsian dinosaur—the group that contained such horned dinosaurs as Triceratops, Styracosaurus, and Monoclonius. Any dino buff cannot fail to have noticed the glaring errors that make even this identification a nonstarter. Ceratopsians had rounded elephantine feet, not hooves. They also possessed a bony frill about the neck that an observer could not have failed to notice. Horned dinosaurs lacked this odd animal's scaly shoulder hump. Lepage's animal is a complete chimera, and the story sounds like a fabrication.

The unlikely saurian was back in the news on December 4th:

News apparently corroborating the report of the existence in the Congo of a monster known as a Brontosaurus (the thundering saurian) comes from Elizabethville.

A Belgian prospector and big game hunter named M. Gapelle, who has returned from the interior of the Congo, states that he followed up a strange spoor for twelve miles and at length sighted a beast certainly of the rhinoceros order with large scales reaching far down its body. The animal, he says, has a very thick kangaroo-like tail, a horn on its snout, and a hump on its back. M. Gapelle fired some shots at the beast, which threw up its head and disappeared back into a swamp.

The American Smithsonian expedition was in search of the monster referred to above when it met with a serious railway accident, in which several persons were killed...

Needless to say, the Smithsonian Institute did not find this amusing, especially as several of its members had been killed in a railway accident in Africa. This only confirmed the tall tales in the eyes of both the popular press and the general public. The Smithsonian felt that it had to quash such outrageous nonsense, and wrote a letter to the *Times* which was published on January 21:

Sir,

I am authorised to contradict the statement that the members of the Smithsonian African Expedition who proceeded to this territory came here to hunt the brontosaurus. There is no foundation for this statement. I may also state that the report of the brontosaurus arose from a piece of practical joking in the first instance, and, as regards the prospector "Gapelle," this gentleman does not exist except in the imagination of a second practical joker, who ingeniously coined the name from that of Mr L. Le Page.

Yours faithfully,

WENTWORTH D. GREY

Acting Representative of the Smithsonian African expedition in the Katanga

Elizabethville, Jan 21

Another bogus report was printed in the *Rhodesia Herald* on July 15, 1932, in which a Mr. F. Grobler claimed to have knowledge of the existence of a giant lizard known as the Chepekwe. Grobler stated that it had been discovered six months earlier by a German scientist in the swamps of Angola. The reptile fed on hippos and rhinos, and Grobler claimed to have seen a photograph of the monster squatting on a hippo it had just killed. Grobler's gravitas seemed supported, as he claimed to have acted as a guide to the renowned explorer Hans Schomburgk in his expedition into the Dilolo swamps. The major had stated in a lecture the previous year that a tradition of giant reptiles was prevalent in Central Africa.

Shortly after this, a Swedish man, J. C. Johnson—an overseer on a Belgian rubber plantation—wrote to the *Cologne Gazette* enclosing purported photographs of the creature. These—together with his story—found their way into the *Rhodesia Herald*. The lurid tale runs thus:

On February 16 last I went on a shooting trip, accompanied by my gun-bearer. I only had a Winchester for small game, not expecting anything big. At two in the afternoon, I reached the Kassai valley.

No game was in sight. As we were going down to the water, the boy suddenly called out "elephants." It appeared that two giant bulls were almost hidden by the jungle. About fifty yards away from them I saw something incredible—a monster, about sixteen yards in length, with a lizard's head and tail. I closed my eyes and reopened them. There could be no doubt about it, the animal was still there. My boy cowered in the grass whimpering.

I was shaken by hunting-fever. My teeth rattled with fear. Three times I snapped; only one attempt came out well. Suddenly the monster vanished, with a remarkably rapid movement. It took me some time to recover. Alongside me the boy prayed and cried. I lifted him up, pushed him along, and made him follow me home. On the way home we had to traverse a big swamp. Progress was slow, for my limbs were still half-paralysed with fear. There in the swamp, the huge lizard appeared once more, tearing lumps from a dead rhino. It was covered in ooze. I was only twenty-five yards away.

It was simply terrifying. The boy had taken French leave, carrying the rifle with him. At first, I was careful not to stir, then I thought of my camera. I could plainly hear the crunching of rhino bones in the lizard's mouth. Just as I clicked, it jumped into deep water.

The experience was too much for my nervous system. Completely exhausted, I sank down behind the bush that had given me shelter. Blackness reigned before my eyes. The animal's phenomenally rapid motion was the most awe-inspiring thing I had ever seen.

I must have looked like one demented, when I at last regained camp. Metcalf, who is boss there, said I approached him, waving the camera about in a silly way and emitting unintelligible sounds. I dare say I did. For eight days I lay in a fever, unconscious nearly all the time.

It seems the herbivorous Triceratops/Brontosaurus had been transformed into the savage, carnivorous *Tyrannosaurus rex*—all the more challenging to the intrepid. Unfortunately, Johnson's picture did not live up to his story. It is a tawdry fake, showing a Komodo dragon inexpertly superimposed on a dead rhino. So poor is the quality that it would not frighten anyone over the age of five, let alone send a supposedly seasoned hunter into a fear-crazed fever for over a week.

The reader might feel a little disheartened at this point, as all the African stories so far have turned out to be hoaxes. There are, however, two points to note.

Firstly, we have sorted the wheat from the chaff and can now proceed to genuine reports; and secondly, the hybrid animal reported seems to have characteristics of both dinosaur-like creatures reported in Central Africa, as there are two distinct kinds. Moreover, one kind is indeed referred to as Chepekwe in some areas.

Let us take a look at this beast first.

A dishevelled tramp peddling gridirons wandered up the garden path of Ethelreda Lewis's Johannesburg home one day in 1925. Most folk would have shooed such an unwholesome fellow off their property, but Mrs. Lewis, being a kindly soul, invited him in for some refreshments. As it turned out, this was a stroke of luck both for the vagrant and for Lewis, who was a novelist. The old man began to reminisce about his past, and literary immortality for both him and his host was assured.

The tatty old gent of the road was one Alfred Aloysius Smith—or "Trader Horn," as he had been better known. The novelist soon realized she had a veritable gold mine in her living room, and transcribed his stories into a series of bestselling books.

Horn's tale was the stuff of pulp fiction. He was born in Lancashire in 1861 and educated in a strict Roman Catholic school (St. Edward's College, Liverpool). Here he was taught French, Portuguese, and Spanish. This did not suit the young tearaway at all, and he was soon expelled for excessive wildness and for "always being on the roof"!

He took a ship to the West African country of Gabon, and there, aged seventeen, started work for a British trading company—Hatton and Cookson Ltd—buying ivory and rubber and selling various trade goods. This is where his story really takes off. Horn claimed all kinds of fantastic adventures, hunting every known jungle beast, canoeing up unexplored rivers, and generally behaving in a manner befitting a character in a Tarzan novel. After five years of these shenanigans, he came home to Lancashire and married his childhood sweetheart. Soon after, they moved to London and, in an attempt to settle down, Horn became a reporter, then a policeman. These—not exactly sedate—jobs failed to excite him enough, so he joined Buffalo Bill's Wild West Show, and moved to Pittsburgh, Pennsylvania, in the US. Here his wife died, and he was gripped by wanderlust once more, and willingly shipped his two children back to relatives in England.

What he lacked as a father he made up for as a traveler. He roamed the world like "The Wandering Jew," visiting Mexico, Australia, Madagascar, and of course his beloved Africa. Eventually, poverty caught up with him and he became a drop-out, ending up in a Johannesburg doss-house. Shortly after, he met Lewis, and "Dame Fate" smiled on him again. So popular was his life story that it was made into a Hollywood film in 1930 (one wonders if Horn ever saw it, and if so, what he thought). Horn died, and was buried in Whitstable, Kent, in 1931.

The obvious question is how much—if any—of Horn's narrative can be trusted. We must remember that he was an old man recalling events of half a century or more before. Also, a warm meal and a roof over his head would have been incentive enough for him to spin the wildest yarns for his host's entertainment. Finally, Lewis herself probably spiced up the stories with a novelist's style.

Perhaps we should not be so quick to reject all of Horn's adventures—some quite reputable persons have held stock in what he said. One such was Dr. Albert Schweitzer, who commented, "Apart from a few unimportant slips, the statements made by Trader Horn about the country are generally accurate."

It would be surprising if Horn had not heard of "dinosaurs" in Africa, and true to his reputation, he does not disappoint us. Once, by some lakes in the Cameroons, he came across a three-toed footprint as large as a frying pan. This he linked to a creature known as the Amali, which was spoken of by pygmy bushmen. He also claimed to have seen carvings of it in their caves. This curious track turns up again in the saga of the Africa monsters.

Carl Hagenbeck believed in a giant saurian haunting the swamps of Africa, but appears to have only known of a "Brontosaurus"-type creature. Some of his informants, however, also knew of a short-necked, horned beast. Hans Schomburgk, for example, had heard tell of a dangerous animal lurking in Lake Bangweulu in East Africa. The animal was said to kill hippos, but malaria prevented Schomburgk investigating further.

It was another English expat who gathered more information on the horned giant of Lake Bangweulu. J. E. Hughes was born in Derbyshire in 1876 and attended Cambridge University. After this, his family apparently expected him to accept a career in the Church of England. This apparently repulsed him so much that—much like Trader Horn before him—he rebelled. The British South Africa Company offered him a job as Assistant Native Commissioner in the newly formed civil service of northeast Rhodesia. After seven years of service, Hughes resigned and became a hunter/trader. He lived for the next eighteen years on the Mbawala islands on Lake Bangweulu. He recorded his life in a book, *Eighteen Years on Lake Bangweulu*, in which he writes of the monster:

> *For many years now there has been a persistent rumour that a huge prehistoric animal was to be found in the waters of our Lake Bangweulu. Certainly the natives talk about such a beast, and "Chipekwe," or "Chimpekwe," is the name by which they call it.*

> *I find it is a fact that Herr Hagenbeck sent up an expedition in search of this animal, but none of them ever reached the Luapula or the lake, owing to fever, etc.; they had come at the wrong time of year for newcomers.*

> *Mr. H. Croad, the retired magistrate, is inclined to think there is something to the legend. He told me one night, camped at the edge of a very deep small lake, he heard a tremendous splashing during the night, and in the morning found a spoor on the bank not that of any animal he knew, and he knows them all.*

> *Another bit of evidence about it is the story Kanyeshia, son of Mieri-Mieri, the Waushi Paramount Chief, told me. His grandfather had said that he could remember one of these animals being killed in the Luapula in deep water below the Lubwe.*

A good description of the hunt has been handed down by tradition. It took many of their best hunters the whole day spearing it with their "Viwingo" harpoons—the same as they use for the hippo. It is described as having a smooth dark body, without bristles, and armed with a single smooth white horn fixed like the horn of a rhinoceros, but composed of smooth white ivory, very highly polished. It is a pity they did not keep it, as I would have given them anything they liked for it.

I noticed in Carl Hagenbeck's book Beasts and Men (abridged edition, 1909, p. 96) that the Chipekwe has been illustrated in bushman paintings. This is a very interesting point, which seems to confirm the native legend of the existence of such a beast.

Lake Young is named on the map after its discoverer, Mr. Robert Young, formerly [Native Commissioner] in charge of Chinsali. The native name of the lake is "Shiwangandu." When exploring this part in the earliest days of the Administration, he took a shot at an object in some floating sudd[sic] that looked like a duck: it dived and went away, leaving a wake like a screw steamer. This lake is drained by the Manshya river, which runs into the Chambezi. The lake itself is just half-way between Mipka and Chinsale Station.

Mr .Young told me that the natives once pulled their canoes up the Manshya into this lake. There were a party of men, women, and children out on a hippo-harpooning expedition. The natives claimed that the Guardian Spirit of the lake objected to this and showed his anger by upsetting and destroying all the men and canoes. The women and children who had remained on the shore all saw this take place. Not a single man returned, and the women and children returned alone to the Chambezi. He further said that never since has a canoe been seen on Lake Young. It is true I never saw one there myself. Young thinks the Chipekwe is still surviving there.

Another bit of hearsay evidence was given me by Mr. Croad. This was told to him by Mr. R. M. Green, who many years ago built his lonely hermitage on our Lulimala in the Ilala country about 1906. Green said that the natives reported a hippo killed by a Chipekwe in the Lukulu—the next river. The throat was torn out.

I have been to the Lukulu many times and explored it from its source via the Lavusi Mountain to where it loses its self in the reeds of the big swamp, without finding the slightest sign of any such survival of prehistoric ages.

When I first heard about this animal, I circulated the news that I would give a reward of either five pounds or a bale of cloth in return for any evidence, such as a bone, a horn, a scrap of hide, or a spoor, that such an animal might possibly exist. For about fifteen years I had native buyers traversing every waterway and picking up other skins for me. No trace of the Chipekwe was ever produced; the reward is still unclaimed.

My own theory is that such an animal did really exist, but is now extinct. Probably disappearing when the Luapula cut its way to a lower level—thus reducing the level of the previously existing big lake, which is shown by the pebbled foothills of the far distant mountains.

Perhaps, if we are to believe Mr. Young's tale, the creature's ferocity kept it from being hunted very often. A picture is emerging of a huge, dangerous, semi-aquatic animal with a single horn and an antipathy toward hippos. Many have come to the conclusion that these are ceratopsian dinosaurs. These were a sub-order of Ornithischia (bird-hipped dinosaurs) which contained such well-known horned dinosaurs as Triceratops and Styracosaurus. They were all herbivores, and were typified by bearing horns and a bony frill—like an Elizabethan ruff—that grew from the rear of the skull to protect the animal's neck. The number of horns varied between the species—some, such as Centrosaurs, bore only one horn on the snout.

There are two main stumbling blocks with the dinosaur theory. First and foremost, there is no fossil evidence for any species of non-avian dinosaur surviving beyond the Cretaceous period (which ended sixty-five million years ago). Secondly, there is no indication of any species of ceratopsian dinosaur being aquatic. So we need to look elsewhere for this beast's identity. Let us examine some more evidence.

The *Daily Mail*'s dinosaur fiasco did produce at least one seemingly genuine piece of evidence in what seems to be an honest letter from C. G. James—a gentleman who had resided in Africa for eighteen years. His letter was published on December 26, 1919.

> *Sir, I should like to record a common native belief in the existence of a creature supposed to inhabit huge swamps on the borders of the Katanga district of the Belgian Congo—the Bangweulu, Mweru, the Kafue swamps. The detailed descriptions of this creature vary, possibly through exaggerations, but they all agree on the following points:*

> *It is named the Chipekwe; it is of enormous size; it kills hippopotami (there is no evidence to show it eats them, rather the contrary); it inhabits the deep swamps; its spoor (trail) is similar to a hippo's in shape; it is armed with one huge tusk of ivory.*

It is useful at this point to realize that Lakes Bangweulu and Mweru are connected via the Luapula river system (where supposedly a specimen was killed).

Identical reports have come in from elsewhere in the "dark continent." Lucien Blancou, chief game inspector in French Equatorial Africa, collected stories of unknown animals between 1949 and 1953. Some of these seem to refer to an animal like the Chipekwe.

The Africans in the north of the Kelle district, especially the pygmies, know of a forest animal larger than a buffalo, almost as large as an elephant, but which is not a hippopotamus. Its tracks are only seen at long intervals, but they fear it more than any other dangerous animal. The sketch of its footprint which they drew for M. Millet is that of a rhinoceros. On the other hand, they do not seem to have said that it has a horn, though they have certainly not said that it has not. While M. Millet was at Kelle, in 1950 if I am not mistaken, one of the best-known African chiefs in the district came several days march to inform him that "the beast had reappeared." Unfortunately, this is all I can say, for M. Millet left the district in 1951, and I have not been able to go there myself. The rewards in kind which this official offered the pygmies for tangible proof of the animal's presence yielded no result.

Around Ouesso, the natives talk of a big animal which does have a horn on its nose—though I don't know whether it has one or several. They are just as afraid of it as the Kelle people.

Around Epena, Impfondo, and Dongou, the presence of a beast which sometimes disembowels elephants is also known, but it does not seem to be as prevalent there now as in the preceding districts. A specimen was supposed to have been killed twenty years ago at Dongou, but on the left bank of the Ubangi and in the Belgian Congo.

This report is particularly interesting, as the man in question recognized the print as being that of a rhinoceros—one of the few animals capable of killing an adult hippo. (The hippopotamus is one of the most dangerous animals in Africa. Despite the cuddly Disney image, this animal is in reality totally unpredictable and highly territorial. It also possesses a huge mouth armed with immense, curving tusks, that can bite a man in two, or rend a boat asunder.) In the Congo, this horned animal is called Emela-ntouka. This translates as "killer of elephants." Places where both hippos and elephants are scarce or absent are reputed haunts of this aggressive creature, which gores the former animals to death with its horn.

Ilse von Nolde spent ten years in eastern Angola, and in 1930 reported events much like the ones related previously. Natives told her of a monster called "coje ya menia" or "water lion." The name seemed to relate to the roaring sound the animal produced, rather than to any resemblance to a lion. She heard its rumbling cry for herself on several occasions. It was said to inhabit the water, but was also seen on the bank. In the rainy season, when the Cuanza River was in flood, it moved to smaller rivers and swamps.

One day, she met a native in hippopotamus-skin sandals. She asked him if he had killed the hippo himself, and he replied that he had found the animal dead—killed by a coje ya menia. On another occasion, a Portuguese lorry driver told her of how he had heard of one of these creatures killing a hippo on the previous night. He intrepidly set off to investigate with several native hunters and found the beast's tracks. The hippo's tracks ran for several miles and seemed intermingled with the tracks of its pursuer—but none of them could identify them. Finally, they came upon an area where the grass and bushes had been smashed and crushed. The mangled cadaver of the hippo lay in the centre of the devastation. It looked as if it had been hacked and ripped by a huge bush knife. None of the carcass had been eaten. It would seem that the only thing capable of inflicting such wounds would have been a massive horn.

For me, the clinch in this animal's identity is a photograph taken in 1966 in the Congo by French photographer and naturalist Atelier Yvan Ridel. The photo shows a large three-toed footprint—one of a set that led out of a mass of reeds, up a steep bank, across a small beach, and into the river.

The tracks are instantaneously recognizable to any zoologist worth his salt—they are the footprints of a rhinoceros. The nearest rhino populations to the Congo are a thousand miles away in the Cameroons and the Central African Republic. These are black rhino (*Diceros bicornis*)—the smaller of the two African species and much smaller than the reports of the Emela-ntouka. The toes seem a little more elongated than those of other rhinos, and this may be an adaptation to a marshy environment. The rhino's close relatives in the order Perissodactyla (odd-toed ungulates)—the tapirs—display slightly elongated toes and are invariably found in swampy biotopes.

The Emela-ntouka/Chipekwe is most likely to be, not a ceratopsian dinosaur, but a giant semi-aquatic rhinoceros. The idea of a water-dwelling rhino may seem strange, but the great Indian rhino (*Rhinoceros unicornis*) spends almost as much time in water as a hippopotamus. It feeds mainly on lush water plants, such as reeds and water lilies. The Indian rhino also bears only one horn—much like the Emela-ntouka, and unlike the two savannah-dwelling African species which both have two horns.

This unknown species must be a veritable giant. Natives say that it rivals the elephant in size. The largest known rhino is the African white rhino (*Ceratotherium simum*), which can reach five tons in weight, and is second only to the elephants as the largest land mammal. A white rhino would have no trouble dispatching a hippo, but if the Emela-ntouka does indeed kill elephants, it would need to be even more massive. One prehistoric rhino, Paraceratherium, was the largest land mammal of all time, reaching twenty tonnes in weight—bigger than the largest mammoth. One group of rhinos, the Amynodontids, —specialized in an aquatic lifestyle. These flourished in the Oligocene epoch—thirty-eight to twenty-five million years ago—finally dying out about ten million years ago. Could one species have survived into the present? This is by no means impossible, but it is perhaps more likely that our unknown giant is a modern species that has avoided detection, rather than a prehistoric survivor.

But what of the ivory horn? Rhino horn is made from keratin—a fibrous material that also forms human fingernails—and is very different to ivory. This is the only sticking point with the rhino theory. Could the natives be mistaken on this point? I think the answer has to be "yes." However much we want this creature to be a dinosaur, the bulk of the evidence points toward a giant aquatic rhino.

So it seems that our first horned "dinosaur" is no such thing. But what of the second possible dinosaur that lurks in the African rain forest? This is a very different beast.

Lewanika was a king who ruled over the remnants of the Barotse Empire on the middle of the Zambezi River. Prior to the 1920s, Barotseland lay in the northwest district of what is now Zambia. King Lawanika was fascinated by the animals of his kingdom and studied them in detail. His subjects repeatedly told him of a vast aquatic reptile, larger than an elephant. The king gave orders to be notified immediately the next time such a creature appeared. The following year, three men came to his court and told him they had just seen a monster on the edge of the marshes. They described it as taller than a man, with a snake-like head on a long neck. On seeing the men, it slid on its belly into the deep water. The king rode at once to the spot, and saw the depression made by the creature, and the channel where it had slid into the swamp. He told the British Resident—Colonel Hardinge—that the channel was "as large as a full-sized wagon from which the wheels had been removed."

The wagons used by the Boers at the time were four feet, five inches wide, so whatever made the track was a substantial animal. Hardinge found out that the natives called the creature Isiququmadevu.

The description given by the king's three subjects is one we see time and time again across sub-Saharan Africa. An elephant-sized beast, with an elongated neck terminating in a small head, a barrel-shaped body with four sturdy legs, and a long, whip-like tail. The description is reminiscent of a group of saurischian (lizard-hipped), herbivorous dinosaurs called sauropods. These included such well-known dinosaurs as Diplodocus, Brontosaurus, Apatasaurus, and Brachiosaurus. One species—*Amphicoelias fragilimus*—may have been the largest animal that ever lived. At possibly two hundred feet-plus in length, and 230 tons or more in weight, it would have dwarfed even the blue whale. Many kinds of sauropods were found in the African continent, such as Vulcanodon and Aegyptosaurus. The jungles of Central Africa have remained largely unchanged since the Cretaceous period, and many believe that they still harbour dinosaurs.

The first Western involvement with these creatures came in 1913, when the Likuala-Kongo German expedition penetrated the northern Congo. It was led by Freiherr von Stien zu Lausnitz, a colonial officer. The endeavour was due to last two years, but was cut short by the outbreak of the First World War. Lausnitz's report was never published, but parts of the manuscript were obtained by pioneering cryptozoologist Willy Ley. Ley discovered that, during their travels, the Germans collected reports of a giant aquatic animal much feared by the natives. It haunted the lower Ubangi, Sanga, and Ikelemba rivers and was known as Mokele-mbembe. Lausnitz said that the characteristics of the animal had been repeated to him by several experienced native guides who had no knowledge of each other.

The creature is reported not to live in the smaller rivers, like the two Likulalas, and in the rivers mentioned only a few individuals are said to exist. At the time of our expedition a specimen was reported from the non-navigable part of the Sanga River, somewhere between the two rivers Mbaio and Pikunda, unfortunately in a part of the river that could not be explored due to the brusque end of our expedition. We also heard about the alleged animal at the Ssombo river. The narratives of the natives result in a general description that runs as follows:

The animal is said to be of a brownish-grey colour with a smooth skin; its size approximating to that of an elephant; at least that of a hippopotamus. It is said to have a long and very flexible neck and only one tooth but a very long one; some say it is a horn. A few spoke about a muscular tail like that of an alligator. Canoes coming near it are said to be doomed; the animal is said to attack the vessels at once and to kill the crews but without eating the bodies. The creature is said to live in caves that have been washed out by the river in the clay of its shores at sharp bends. It is said to climb ashore even at daytime in search of food; its diet is said to be entirely vegetable. This feature disagrees with a possible explanation as a myth. The preferred plant was shown to me; it is a kind of liana with large white blossoms, with a milky sap and apple-like fruits.

A sauropod with a horn? One sauropod, Amargasaurus, sported a crest of spines on its upper neck. These were believed to have been used for defence against contemporary predators, and to have been rattled—like porcupine spines—for communication and warning. These seem very different to the single horn spoken of here. The great Belgian cryptozoologist Bernard Heuvelmans has suggested that some confusion exists between the

Mokele-mbembe and the Emela-ntouka. Both are large aquatic animals, both are rarely seen, and both now probably exist only in the most remote reaches of the Central African rain forest. It seems that some of their characteristics have been transposed upon each other. The horn of the Emela-ntouka is erroneously placed on the Mokele-mbembe, whilst the long tail of the latter is sometimes attached to the former—a feature never found in rhinos. It should be noted that these confusions only occur in a small minority of reports.

Having said this, there is a possibility that the Mokele-mbembe does sport a horn, albeit one much smaller than that of the Emela-ntouka. Air security officer A. S. Arrey told cryptozoologist Philip Averbuck of a sighting of two long-necked monsters he had as a child. Arrey was living at Kumba in Cameroon during 1948–49. He was swimming with friends in Lake Barombi Mbo. Also present were some British soldiers. Something began to make the waters at the centre of the lake boil, and everyone swam to shore. Looking back, they saw a long neck—some twelve to fifteen feet high—break the surface. A few minutes later, a second, slightly smaller neck rose around two hundred yards from the first. Both were covered by smooth scales. The larger of the two—which Arrey assumed to be the male—bore an eight-inch horn on its two-foot-long head. The smaller creature was hornless. The monsters remained at the surface for an hour before the "female" submerged, followed soon after by the "male."

Lucien Blancou—whom we met earlier—was also aware of this second kind of water monster. In the 1930s, the Linda Banda people of French Equatorial Africa described to him an animal they called "Ngakoula-ngou." It was a gigantic snake-like animal that killed hippos without leaving any sign of a wound and browsed on trees without leaving the water. "Snake-like" seems to refer to the creature's neck, as no true snakes eat vegetation.

Blancou was told by Yetomane—a chief and hunter of great renown—that in 1928, one of the monsters had crushed a field of manioc belonging to the chief, and left tracks 1.90 metres wide. This probably meant the track left by the animal's body rather than individual footprints. The size is quite comparable to that left in the marshes of Barotseland. The same animal was said to have killed a hippo (what is it with African water monsters and hippos?) in the River Brouchouchou. The corpse was eaten by the villagers.

The Baya people told Blancou of an identical creature they knew as badigui. A man called Moussa related how he had seen the beast in his youth:

When he was about fourteen years old and the whites had not yet come (probably around 1880), Moussa was out laying fish-traps with his father in the Kibi stream, which runs into a tributary of the Ouaka called the Gounda in what is now the Bakala district. It was one o'clock in the afternoon in the middle of the rainy season. Suddenly Moussa saw the badigui eating the large leaves of the roro, a tree which grows in forest galleries. Its head was flat and a bit larger than a python's (Moussa spread his hands and put them together to show me the size). Its neck was as thick as a man's thigh and about 4.50 metres long, much longer than a giraffe's; it had no hair but was as smooth as a snake, with similar markings. The underneath of its neck was lighter—also like a snake's. Moussa did not see its body.

His father told him to follow him and run away. The animal gave no cry, but Moussa had heard its cry at other times when he had not seen the beast himself. He did not imitate it for me.

According to him, the old men believed the badigui does not frequent places where you find hippopotami, for it kills them.

Along the Cavally River in Liberia—close to the border with the Ivory Coast—there are similar stories. This river rises in the Niam Mountains in a still-unexplored area. At a place known as Juju Rock, a grey dinosaur-shaped animal is said to live. Natives say it is larger than a crocodile and carnivorous. The diet here seems to be at odds with reports from other areas of Africa, but as we shall see later, there may be an explanation for this. To my knowledge, this creature has never been investigated by Western science, and due to Liberia's current unstable climate, it will probably remain so.

Jorgen Birket-Smith of the Institute of Comparative Anatomy at the University of Copenhagen was a resident in the French Cameroons during the winter of 1949–1950. He was based at Case du Nyong along the River Nyong. He was told by two old hunters of a large animal that inhabited the River Sanaga. He was told that it was larger than a crocodile or hippo, and that it had a long neck like a giraffe. Smith, who had heard rumours of African dinosaurs, drew a picture of a Brontosaurus. The natives instantly identified it as the creature:

> ...the guard remembered, that when he was a boy—that must have been sometime in the twenties—they caught one in the village. Its size was between a hippo and an elephant. The whole village had eaten from it for a week...

> ...it ate from the trees, by which they meant the trees in the gallery forest leaning over the water or turned over into the water. It hardly ever came onto dry land. It is supposed to browse mainly at night, and it stays submerged during daytime.

James Powell, member of the crocodile specialist group of the International Union for Conservation of Nature and Natural Resources, was studying crocodiles in Gabon, along the Ogowe and N'Gounie rivers, when he heard stories of a dinosaur-like animal. The Fang people—former cannibals who had been gradually migrating toward Africa's west coast from inland for some two hundred years—told of a monster they named N'yamala.

Powell befriended a Swiss dentist from the Albert Schweitzer hospital who had married a Fang girl. He accompanied the dentist up the Ogowe to his wife's village. There he became aquatinted with the village witch doctor, Michel Obiang, and found the septuagenarian to be highly intelligent. Powell showed the old man pictures of Gabonese animals such as leopards, gorillas, crocodiles, and hippos. The witch doctor identified them all. Powell then showed him a picture of a bear, an animal not native to sub-Saharan Africa. This was unknown to him. Then he was presented with a picture of Diplodocus, a sauropod dinosaur. He answered, matter-of-factly, "N'yamala." He added that this animal fed on jungle chocolate, a plant with nut-like fruits that grows near riverbanks and lakes. This recalls Lausnitz's description of the Mokele-mbembe's food source half a century earlier.

The witch doctor was insistent that the N'yamala had no horn and rejected pictures of other dinosaurs as unknown to him. A Pterodactyl was not unreasonably identified as a bat.

The following day, Powell travelled eighty miles downstream to study the narrow-snouted crocodile (*Crocodylus cataphractus*). He repeated his experiment with the population of a small village. The results were the same—Diplodocus was instantly identified as N'yamala. The villagers told him that it lived in remote lakes in the jungle. None of them had seen it personally. The area was sparsely populated, and according to the American Embassy in the country, Gabon is still 80 percent unexplored.

Powell made a return to the witch doctor's village at a later date and talked to him again. This time Obiang told him of his own encounter with the N'yamala. In 1946 he had been halfway up the River N'Gounie, where the Ikoy tributary branches off. He had camped for several days by a small cove off the main river, where the waters were deeper. He observed the monster leaving the water at around five in the morning and feeding on jungle chocolate. It was around thirty-three feet long, and as heavy as an elephant. Obiang said it was as strong as one of the caterpillar tractors used at the construction of the hospital. It had thread-like filaments running down its neck, and two "pouches" in the vicinity of the front-legs. These—Obiang stated—were used for storing food, much like a hamster's cheek pouches. This may seem odd, but sauropod dinosaurs possessed a crop much the same as that which birds have. Because they only had small peg-like teeth for nipping off vegetation and totally inadequate for mastication, they processed their food with the crop. This was a highly muscular section of the throat where fibrous plant matter was crushed into a digestible pulp. Sauropods swallowed stones known as gastroliths for this purpose and stored them in the crop for grinding food (the dinosaur equivalent of dentures!). These stones can still be found today, recognizable due to their highly polished nature.

Powell asked Obiang to take him to the spot where his sighting occurred. The witch doctor obliged, and Powell discovered a remote lake in the dense jungle, swarming with ants and flies. The lake was some hundred feet across and eighteen feet deep. Obiang was very scared that the N'yamala might still be in the area, though nothing was seen. When asked if any hunting trophy, such as a skull, bones, or skin, were ever preserved, Obiang replied, "Oh no, no, the N'yamala is king of the waters. It never dies. No one ever kills a N'yamala."

Obiang suggested that, to see a specimen, Powell should travel down the Ogowe toward the coast. On an island in the middle of a wide part of the river, a N'yamala had killed a hippo, but lack of time prevented Powell from investigating this.

Once home, Powell contacted Paul W. Richards, an authority on the African rain forest.

Richards identified the "jungle chocolate" as a species of Landolphia—a large group belonging to the dogbane family—or Apocyanaceae, which includes vines and lianas.

In January 1980, Powell, together with Dr. Roy P. Mackal—a biochemist from the University of Chicago and vice president of the International Society of Cryptozoology—mounted an expedition to look for surviving dinosaurs in the Congo. Early on in their trip, they met an eyewitness at a mission. Firman Mosomele claimed to have seen the Mokele-mbembe some forty-five years before, at a bend in the Likouala-aux-Herbes river just below the town of Epena. He saw a snake-like head supported on a three-metre (ten-foot) neck break the water. Terrified, he paddled his canoe away, as the beast's two-metre (seven-foot) back surfaced. The animal was a reddish-brown in colour. Mosomele said that, in the Epena district, natives were afraid to go to the riverbank in late afternoon, as this was when the creature came ashore to feed.

The dread which this animal instils in natives is illustrated by the beliefs attached to the Mokele-mbembe by locals. Some of these were related to Mackal and Powell by Marien Ikole, a pygmy from the village of Minganga:

- To tell anyone of a sighting or even talk of the beast you will die; this obviously hampers investigation, and implies that the vast majority of sightings go unreported.

- When the monster appears, it causes a miniature tidal wave that will wash onto the bank, and then suck you back into the water to be drowned.
- The Mokele-mbembe is so huge that it can bridge the river.
- Once, in a time of war, a tribe escaped from their foes by running across the creature's neck, back, and tail, as it positioned itself across the river.

Pascal Moteka—another pygmy who lived near Lake Tele (infamous for sightings)—recounted to the expedition the killing of a Mokele-mbembe shortly before his birth (around 1950). The fishermen were too afraid of going out onto Lake Tele due to the monsters who entered it via certain waterways connected to the swamps, called "molibos." The tribesmen cut down some trees about six inches in diameter and trimmed off the branches. Then they sharpened one end of each and rammed the blunt ends into the mud at the bottom of one of the waterways to form a barrier against the monster. One of the creatures tried to smash its way through, and whilst it was entangled on the spikes, the pygmies managed to spear it to death.

There was a great celebration, and the animal's carcass was butchered and eaten. However, all who ate the flesh of the Mokele-mbembe were poisoned and died soon afterwards. This recalls the almost universal belief in the toxicity of dragon blood. In many British dragon legends, so much as a drop may be lethal, and many victorious heroes met their end by spilling the blood of their terrible foes. This tale was later confirmed by other fishermen in the area.

Pascal had seen the animals himself, mainly in mid-morning. He related seeing their long necks rise seven feet from the water, and on occasion a rounded back surfacing like a buoy. His intense fear of the creatures prevented him from approaching them, and consequently he only observed them from a distance.

Other witnesses saw the animal at much closer range. Nicolas Mondogo said that his father had seen a massive animal with a long neck come out of the river and onto a sandbank. It left dinner-plate-sized tracks and a great furrow where its tail had dragged in the wet sand. This occurred between the villages of Mokengi and Bandeko, on the upper Likouala-aux-Herbes river.

Close to this spot, Nicolas had his own sighting when aged seventeen. It had been seven in the morning, and he was on his way to a Catholic mission at Bandeko. He had paused to hunt some monkeys when a huge animal rose from the river. The water in the area was only a metre (three feet) deep, so he could observe the underbelly and legs of the animal. The beast stayed in view for three minutes. It was reddish-brown in colour, with a long neck (as thick as a man's thigh) and a head that bore a comb like a rooster. It was a mere forty feet away and was some thirty feet long. It stood seven feet high, and possessed a neck of a similar length, giving a height of approximately fourteen feet, quite comparable to that of a giraffe. The tail seemed longer than the neck.

David Mambamlo, a schoolteacher, saw it even closer. Only three years before, he had been canoeing just upstream from Epena at about three in the afternoon when a two-metre (seven-foot) head and neck broke the surface only ten metres (thirty feet) from his vessel. He was mesmerised with horror, as it rose further out of the river, exposing its upper breast. The monster was grey in colour with no visible scales. David picked out a picture of a Brontosaurus from a book and said that the animal most resembled that. He subsequently showed the expedition the location of his sighting, a cave in the riverbank half a mile from the village. The water level had dropped, revealing the cave, but no occupant was spied.

Daniel Omoa, a Ministry of Agriculture worker, said that in July of 1979, a Mokele-mbembe had taken up residence in a pool close to the river. This was just north of Dzeke, fifty miles downstream from Epena. The people saw it leave the jungle and enter the river by a sandbar that had become a small island during the dry season. Elephant-sized tracks were found on the island, and a pathway of crushed grass seven feet wide was found at the riverbank.

Mackal and Powell had to return to the States soon after. Though they had not seen the Mokele-mbembe, both were now convinced of its existence as a rare but real biological entity, as opposed to just a native myth.

In October 1981, Mackal returned to the Congo, this time accompanied by Richard Greenwell from the University of Arizona. Greenwell was also the secretary of the International Society of Cryptozoology. This time they came closer than ever to seeing the Mokele-mbembe.

Whilst travelling upstream from the village of Itanga, they came to an area where the trees briefly thinned out along the riverbank and were replaced by elephant grass. As they rounded a sharp bend, something dived off the five-foot bank and into the water with a great splash. Whatever it was, it was large, as it caused a wash to flow over the expedition's dugouts. Due to the shaded nature of the area, none of them got a good look at the creature. The pygmies were in no doubt of the creature's identity and screamed out "Mokele-mbembe" in terror. The group searched the area for half an hour, but the monster (if that was what it was) did not reappear. The water into which it dived was found to be twenty-three feet deep.

They travelled to Dzeke, but found that the animal dwelling in the nearby pool had departed some sixteen months previously, in June 1980. A villager called Apollinaire related his sighting of the animal. Whilst hunting monkeys, he saw its long head and neck rear up into view and saw it feed on malombo (another name for jungle chocolate).

They were also told that the creatures were once common, before the coming of the white man. When motorboats began to come up the river, they retreated to more remote areas.

At the village of Bozenzo, they were told of a Mokele-mbembe that reached with its neck from the river and seized goats. The creature then devoured them. This seems at odds with the insistence elsewhere that the animal is an herbivore. Village men plucked up the courage to tackle the brute after several livestock losses. They attached ropes to their spear shafts and harpooned the monster. It was cut up and eaten. After this, the village became cursed. Houses burnt down for no apparent reason, illnesses broke out, and there were strange deaths. As a result, the Mokele-mbembe became a venerated animal. This occurred around 1908.

Perhaps this animal has some sort of toxin in its flesh. This is far from impossible. Poisonous birds have recently been discovered in New Guinea, and South America's poison-dart frogs are some of the most toxic animals known to man. Alternatively, food poisoning could arise if the meat of this unknown animal was not properly cooked. Either way, an outbreak of illness and death would be attributed to a magic curse from the creature.

At Dzeke they found a track smashed through the jungle at a height of seven feet, and a trail of foot-long prints. The village folk still feared the pool where the monster had laired and would not approach it. Had they investigated the pool during the first expedition, they might well have found their quarry.

Since then, there have been many attempts to find the Mokele-mbembe. Herman Regusters, an engineer from California, led an expedition into the Congo with his wife in September 1981. Originally Regusters had planned to co-run Mackal's second venture, but then he changed his mind without adequately explaining why. He claimed he saw—and photographed—a Mokele-mbembe, but his shots were hopelessly underexposed and of no value. This sort of bad luck seems to habitually dog cryptozoologists. Much to the delight of sceptics (who are generally armchair zoologists), cameras seem to be absent, or to fail, whenever an unknown animal appears. The next expedition suffered a similar fate.

A search for the monster by Congolese scientists took place in the spring of 1983. Lake Tele was visited by a group from the Ministry of Water and Forests. It was led by Dr. Marcellin Agnagna, a zoologist from the Parc de Zoologie. On May 1, whilst filming monkeys in the forest, Agnagna was approached by an excited local who bade him to come quickly to the lake. Wading into the water, he saw the back and neck of a large animal some seven hundred feet away. The strange animal turned its head as if it had heard Agnagna approach. As he raised his camera and began filming, he realized he could not see anything through the viewfinder. He had foolishly kept the camera on the "macro" setting. In the time that it took him to realize this and switch the setting, his film had run out! Despite this mother of all frustrations, he watched the animal through the telephoto lens and obtained a detailed sighting. The head was held around three feet out of the water and turned from side to side as if listening. It was reddish-brown, with a long, slender muzzle and crocodile-like oval eyes. Just behind the neck was a black hump. Agnagna was sure the beast was a reptile, but not a crocodile, turtle, or python.

He and two particularly brave villagers waded further into the lake. The animal submerged, but then surfaced again, staying in view for twenty minutes. Agnagna took some shots with a small thirty-five-millimetre camera, but the pictures were too indistinct for identification.

British explorer Bill Gibbons led two expeditions. Operation Congo took place in 1986 and was an Anglo-Congolese effort. Despite being hampered by bureaucratic problems from the Congolese side, they rediscovered Mackal's Mokele-mbembe tracks from his second trip, thus independently verifying them. They were also told that Lake Tele itself is not the home of the monsters, but that they live in three sacred lakes in the surrounding jungle. These lakes are holy to the Boha people who claim ownership of Lake Tele. Two expedition members were allowed to spend a day at one of these lakes, never before seen by white men. They were informed that the Mokele-mbembes use Lake Tele as a feeding ground and for transit, but they actually reside in the smaller bodies of water. However, no evidence for their presence in the sacred lakes was uncovered.

Gibbons met up with American journalist Rory Nugent at Epena. Nugent was conducting a solo expedition by boat, plane, and on foot. At Lake Tele he saw a black periscope-shaped object break the surface. It was about a thousand metres away, and beyond the range of his camera. He tried to photograph the object, however, but once again the pictures were far too blurred.

The second Operation Congo expedition took place in late 1992. Though no further evidence was gathered, two unexplored lakes were visited: Lake Tibeke and Lake Fouloukou.

Most recently, English explorer Adam Davies attempted to solve the riddle. Impressively, he self-financed his trip and travelled solo—no mean feat in such an unstable and dangerous country. He was told by Mokele-mbembe witness Dr. Marcellin Agnagna that Lake Makele—some 1.25 miles from Lake Tele—was the best place to look for the beast. The pygmies told him that they believed that only the male of the species bears a horn. The confusion with the horn seems to continue. Sadly, Adam did not catch sight of the monster himself in the limited time he had in the area. Hopefully, one day an expedition will have the luck to prove conclusively just what this most enduring of "neo-dinosaurs" actually is.

The Mokele-mbembe probably exists. The native accounts (save for the horn) are very consistent. But is it a dinosaur?

The answer is almost certainly no.

The two major factors against this idea are the same as those that applied to the Emela-ntouka. We have seen that there are no fossils that suggest any kind of non-avian dinosaur survived the extinctions at the end of the Cretaceous period. Other "prehistoric survivors" such as the okapi (a short-necked giraffe from Central Africa) and the coelacanth (a primitive fish of the "extinct" order Crossopterygii) have fossil precedents for their survival.

Secondly, and perhaps more importantly, as mentioned briefly earlier in this chapter, sauropod dinosaurs (like most dinosaurs in general) were not aquatic. The notion of aquatic dinosaurs had its sorry genesis with Victorian scientists who perceived these giant animals as being too heavy for their legs to support them for very long on dry land. They envisioned sauropods as spending 90 percent of their lives in water, buoyed up to relieve them of their vast weight, and feeding on soft water plants. Only when they needed to lay eggs would they leave this safe environment. (Equally wrong, it was once believed that carnivorous dinosaurs could not swim! In fact, they were the best at swimming of all dinosaurs.)

This ridiculous theory has long been disproven. Sauropod bodies were adapted for life on land, with thick pillar-like legs to support them and hollow vertebrae to reduce their body weight. Their lifestyle could be compared to that of modern elephants or giraffes: huge herd-dwelling herbivores that browse vegetation. Fossil trackways show sauropods to have been social animals, and that the young travelled with the herd and were not abandoned in lakeside nests as once thought. Lastly, it was physically impossible for sauropods to have been aquatic, on account of their long necks. Previously they were pictured as strolling along lake beds with only their heads above the water. Some, like Brachiosaurus, had their nostrils on top of the head on an elevated crest. Again, this was thought to be an adaptation to aquatic living, a type of "snorkel." However, the water pressure on a sauropod's neck would have caused it to cave in, and if it tried to take a breath, its lungs would have collapsed. Obviously, sauropods entered water from time to time, as do modern elephants, in order to bathe, cool off, and rid themselves of parasites, but they were not habitual water-dwellers.

The Mokele-mbembe does seem to be genuinely aquatic, and hence is unlikely to be a dinosaur.

If it is not a dinosaur, then what is the Mokele-mbembe? The creature that most fits the picture is a giant monitor lizard. In the last volume we encountered Australia's *Megalania prisca*. If giant monitors can exist in the Antipodes, then why not elsewhere? The African animal seems radically different to Megalania, but this may be due to its aquatic habitat.

Many monitors favour a semi-aquatic existence. Most of these amphibious species—such as the water monitor (*Varanus salvator*)—have elongated necks. Monitors are primarily carnivores, but are very adaptable. It is not out of the question that an omnivorous variety could have evolved. One must remember the stories of the Mokele-mbembe eating goats; this lends weight to the omnivore hypothesis. Dr. Ralph Molnar—an expert on Australian fossil monitors—believes that some of them were aquatic and bore crests on their heads. This instantly brings to mind the native descriptions of the Mokele-mbembe.

A thirty-foot-long aquatic, omnivorous monitor seems much more likely than a sauropod species having survived for sixty-five million years.

Giant Ground Sloths

Today's sloths are small creatures, tipping the scales at a maximum of seventeen pounds. The six living species are all found in the neo-tropics and spend most of their time hanging upside down from trees. However, in prehistoric times, there were several families of large sloths that lived entirely on the ground. They lived from around thirty-five million years ago, with the last known species, island dwellers, becoming extinct around four thousand years ago. The largest species, *Megatherium americanum*, were as large as elephants. Most others ranged from human-sized to as large as a brown bear (200 to 1,000 pounds). Ground sloths moved on all fours but could rear up onto their hind legs. They were armed with huge claws for ripping down vegetation and defending themselves from predators. They evolved in South America, and when the Isthmus of Panama formed

2.8 million years ago, they crossed over and colonized North America. A successful group of animals, they lived from the tip of Terra del Fuego in South America up into Canada.

Ground sloths mostly vanished at the end of the last ice age around 11,000–10,000 years ago. It is unknown if climate change, human hunting, or a mixture of both led to their demise. However, there are those who believe that some of the smaller species of ground sloth may survive in the present day.

The bones and preserved hides of such creatures had been unearthed in the nineteenth century. Professor Florentino Ameghino of the Buenos Aires Museum had received such a hide. Though it looked comparatively fresh, it was over 11,000 years old, having been preserved in a cave. However, one of the professor's friends—Ramon Lista, the Argentine Secretary of State—may have seen a live one in 1898.

Lista was on an expedition into the heart of the province of Santa Cruz in southern Patagonia. He and his companions saw a quadrupedal animal that was shaped like a pangolin. Pangolins, sometimes called scaly anteaters, are not closely related to the true anteaters of South and Central America. They are found only in tropical Africa and Asia. Pangolins eat ants and termites and have powerful claws to rip into termite mounds. They can move on all fours or on their hind legs, and have a long tail. They are covered in scales made of keratin and have a long tail. The beast Lista and company saw had no scales, but was covered in reddish-grey hair. The men shot at it, but it was apparently unhurt and vanished into the bush.

The overall shape of a pangolin recalls a ground sloth, except in size. Both have a heavy tail, claws, and the ability to move on both two and four legs.

The local Tehuelche Indians spoke of an ox-sized beast that was armed with hooked claws. It dug burrows and only emerged at night. Apparently, its hide was proof against bullets.

Indians in the jungles of Brazil speak of a monster called mapinguary. It is covered in shaggy red hair and can stand over seven feet tall. It is armed with hooked claws that can disembowel a man. The mapinguary emits a vile stench and has a terrible roar. The beast is an herbivore and eats bacaba palm hearts and berries. It twists palm trees to the ground to get the palm hearts and travels with herds of white-lipped peccaries (*Tayassu pecari*).

A man named Inocêncio was with ten friends on an expedition up the Rio Uatumã in Pará State, Brazil, in 1930 when he was separated from them and got lost. As he slept in a tree for the night, he heard loud cries coming from a thickset, black figure that stood upright like a man. He shot at it several times and apparently hit it, as there was a trail of blood below his tree.

In 1975, mine worker Mário Pereira de Souza claims he encountered a Mapinguari at a mining camp along the Rio Jamauchím south of Itaituba, Pará State, Brazil. He heard a scream and saw the creature coming toward him on its hind legs. It seemed unsteady and emitted a terrible stench.

American ornithologist Dr. David C. Oren had heard native stories of the Mapinguari and became interested. Oren, a staff scientist at the Goeldi Natural History Museum in Belem, Brazil, conducted more than a hundred interviews in the last nine years with Indians, rubber tappers, and miners who told of having had contacts with the creature.

Oren interviewed seven hunters who claimed to have shot specimens. One group of Kanamarí Indians living in the Rio Juruá Valley claimed to have raised two infant Mapinguaris on bananas and milk; after one or two years, the creatures' stench became unbearable, and they were released. In the late 1990s, Dutch zoologist Marc van Roosmalen heard that people in one village along the Rio Purus, Amazonas State, Brazil, moved their homes across the river after Mapinguari tracks were found nearby.

The natives believe that the beast's hide can turn back arrows and bullets. It can only be killed by shooting it in the head.

One man Oren spoke with was Geovaldo Karitian, a twenty-seven-year-old hunter from the Karitiana tribe. The encounter occurred in 2004, when he was hunting in the jungle near an area that his tribe calls the cave of the mapinguary.

> *"It was coming toward the village and was making a big noise. It stopped when it got near me, and that's when the bad smell made me dizzy and tired. I fainted, and when I came to, the mapinguary was gone."*

His father Lucas confirmed his son's account. He said that when his son took him back to the site of the encounter, he saw a cleared pathway where the creature had departed, as if a boulder had rolled through and knocked down all the trees and vines.

Another witness was seventy-year-old Joao Batista Azevedo, who says he saw a Mapinguari twenty years ago after a forty-five-day canoe ride from the nearest village.

"I was working by the river when I heard a scream, a horrible scream. Suddenly something looking like a man came out of the forest, all covered in hair. He was walking on two legs and thank God he did not come toward us. I will always remember that day."

Another scientist, American ethnobiologist and anthropologist Glenn Shepard Jr., was doing research on local wildlife among the Machiguenga people of the far western Amazon, in Peru. Tribal members all mentioned a fearsome sloth-like creature that inhabited a hilly, forested area in their territory. A member of the tribe remarked matter-of-factly that he had also seen a mapinguary at the natural history museum in Lima. Dr. Shepard checked; the museum has a diorama with a model of the giant prehistoric ground sloth.

In the trackless jungles of South America, you could hide almost anything. The known species of ground sloth seem to have lived in more open grasslands. However, the Mapinguari may be an undiscovered species adapted to deep rain forest, or a species known from the fossil record that adapted to jungle life, perhaps to avoid man. It is also interesting to note that the mylodont genus of ground sloths had osteoderms studding their hide. These are nodules of bone that act as armour. The mylodont has, in effect, a hide like chain mail. The mapinguary is said to be invulnerable to arrows or bullets save for the head. A ground sloth with osteoderms would indeed be hard to kill.

In the first volume, we saw many creatures that were rediscovered after being thought to be extinct. How many more are out there?

CHAPTER TWO

Giant Creatures

"And presently, 'Jab!' a fresh supply of the
Food of the Gods was let loose to wreak
its powers of giantry upon the world."

—Food of the Gods, H. G. Wells

Giant Crocodiles

Man has a habit of exaggeration, which is how many "monsters" are created. We have always wanted the animals who share our planet to be larger, and fiercer, and smarter, than they actually are. The more dangerous we perceive the animal to be, then the greater the tales are that we weave around it. One only has to look at the world of the cinema: *Jaws* (1975), *Grizzly* (1976), *Alligator* (1980), and *Anaconda* (1997), the latter being a tale which bears about as much relationship to reality as do the books of Erik von Daniken! It is not so much the size of the snake that is a problem, but that it was portrayed as a cheetah-fast animal that eats four or five members of the cast in quick succession.

Few carnivores are as dangerous to man as many people would like to think. It is, however, fitting that reptiles are one of the few groups of animals with a real tendency toward gigantism, and that the crocodile is one of the few genuine man-eaters.

To be fair, only two of the twenty-six currently recognized extant species of crocodilian regularly attack humans. These are the Nile crocodile (*Crocodylus niloticus*) and the Indo-Pacific crocodile (*C. porosus*). But these account for five thousand human deaths a year, far more than all attacks by shark, big cat, and bear combined. Of vertebrates, only humans themselves kill more.

How does one define a "giant" crocodile? This is debatable, but for now I am classing any animal of twenty-three-feet or more as a giant! As we shall see, there are reports that suggest that twenty-three-foot specimens may be dwarfed beside the true giants.

There has been a precedent for giant crocodiles in the past. *Deinosuchus riograndensis* was a giant alligator of the late Cretaceous, which flourished in both eastern and western North America. Although it is known only from fragments of its massive skull, estimates of its length are around forty feet.

Sarcosuchus imperator was a giant gharial of North Africa from the early Cretaceous period. Like modern gharials, this species had elongated jaws and specialized in feeding on fish. It attained a length of thirty-six feet. The teeth alone were six inches long, and the armoured dermal scutes that adorned its skin in life measured thirteen by six inches across.

These nightmarish brutes were not confined to the Mesozoic era. In the Tertiary period, a giant caiman (*Purussosaurus brasiliensis*) flourished in the Amazon and grew to forty feet. This was only eight million years ago in the late Miocene, long after the extinction of non-avian dinosaurs.

More recently still—in the Pleistocene epoch—Australia played host to a thirty-foot terrestrial crocodile called *Quinkana fortirostrum*. With teeth as impressive as those of a *Tyrannosaurus rex*, this predator lived on land rather than in the water. Its claws were short and hoof-like, and it is believed that this reptilian giant galloped after its prey!

In the last century or so, travellers have brought back tales of monster crocodiles still lingering in ill-explored corners of the tropics. It is my belief that these are not survivors from prehistory, but vast specimens of animals from known species that far exceed the size limits dreamed of by most zoologists.

Herein I will examine the evidence for giant crocodiles and try to suggest what makes some specimens grow so frighteningly huge.

The Old World seems to have the monopoly on giant crocodilians. Only two New World crocodilians are accorded lengths of seven metres by some authorities. These are the American crocodile (*C. acutus*), an animal which lives in Central America and the tip of Florida, and the Orinoco crocodile (*C. intermedius*), of Colombia and Venezuela. However, very few details have ever been given for either species. The explorer Alexander von Humboldt (1769–1859) claimed to have shot at a twenty-two-foot Orinoco crocodile in Venezuela, but we only have his word for it. The maximum length for either species seems to be only about seventeen feet.

It should be noted that the Indian gharial (*Gavialis gangetius*) has also been credited with huge size. Lorenz Hagenbeck—son of Karl Hagenbeck, world-famous animal dealer—cited one such account. One of his friends was said to have shot a thirty-foot gharial whose bloated carcass looked like that of a stranded whale! However, reports of such giants of this species seem unknown today.

The claims of giant crocodiles from the Old World seem much more believable, if only because of the sheer number of them, and the fact that some were measured by recognized experts.

Officially, the world's largest reptile is the Indo-Pacific crocodile (*Crocodylus porosus*). This animal is also known as the "saltwater" or "estuarine" crocodile. However, it is confined neither to salt water, nor to estuaries. Hence my use of the name Indo-Pacific, which is much more suitable because it defines its geographical distribution.

It is a formidable predator, quite capable of tackling such mighty prey as tigers, water buffalo, and sharks. Man is also firmly on the menu of this awesome beast.

An example of just how powerful the Indo-Pacific crocodile is occurred in the East Alligator River in Northern Australia during March 1987. A Toyota utility truck was crossing the river in about a metre of water. Halfway across the river, the truck met a crocodile. The reptile lashed out with its tail, smashing the truck over onto its side like a toy. The panicked passengers scrambled onto the side of the crippled machine as the crocodile circled it. Finally, some Kakadu National Park rangers came to their rescue in motorboats.

Even more extreme is the case from Princess Charlotte Bay (off the coast of northern Queensland). On the night of June 2, 2000, some people on board a trawler in the bay were awakened by strange sounds. By torchlight they saw a large Indo-Pacific crocodile trying to mate with one of the floats of a seaplane tethered to the trawler. The amorous reptile overturned and sank the aircraft! Needless to say, no one was too keen on a salvage operation.

The biting power of a large crocodile has been measured at ten thousand newtons. This gives it the most powerful bite of any living animal. In the earth's history, only one creature held a more formidable bite—the *Tyrannosaurus rex*!

The largest specimen generally accepted by experts was a twenty-eight-foot, four-inch male shot on the MacArthur Bank of the Norman River, Queensland, Australia, in 1957 by Mrs. Kris Pawloski. The mammoth body was too big to move, but it was photographed, although, sadly, the photograph was lost in 1968. However, her husband Ron was a recognized expert on crocodiles and had carefully measured the specimen. He was astounded at its size, having previously measured no fewer than 10,287 specimens and having found none larger than eighteen feet.

This giant was never weighed (because of the impossibility of moving its immense carcass), but conservative estimates put it at a weight of two tons. Other estimates have produced a figure of three tons or more!

There have, however, been some cases of exaggeration with this species. Bellaire, a British herpetologist, worked out the head-to-body ratio of a typical crocodile (as opposed to specialized species like the gharial) at 1:7.5. This had connotations for the remains of some supposed "giants."

The skull from a supposedly twenty-nine-foot specimen, killed after a six-hour struggle in the Philippines in 1823 by Paul De La Gironiere and George Russell, was measured at twenty-six inches. This gave the true size of the animal as being just under twenty feet. The original skull measurement was given as thirty-six inches. No one knows whether this is a misprint, an exaggeration, or even if the original skull has been lost and replaced with that from a smaller animal.

Another infamous giant cut down to size by Bellaire was a claimed thirty-three-foot crocodile that had been harpooned in the Bay of Bengal in 1940. The skull was measured at twenty-eight inches, and the total length of the animal was therefore given as only twenty-one feet. It transpired later that the wrong skull had been measured, and that the skull of the Bay of Bengal crocodile was much smaller, implying that its owner was not even twenty-one feet in length, let alone the thirty-three feet claimed for it!

Persistent reports, however, argue that—exaggerations aside—Indo-Pacific crocodiles can and do reach massive sizes.

The most famous of these is the creature witnessed in the 1950s by rubber plantation owner James Montgomery. Montgomery's plantation was near the Segama River in North Borneo. He claims to have shot twenty specimens between twenty and twenty-six feet to ensure the safety of workers who washed their laundry at the river.

One particular crocodile dwarfed even these. The local Seluka tribe believed that it was "the Father of the Devil" and threw silver coins into the water to appease it, bringing to mind the dragon-hoards of legend.

Investigating, Montgomery found the beast in question hauled out onto a sandbank. The crocodile filled the whole bank and had the end of its tail in the water. Wisely deciding to leave the monster well alone, Montgomery retreated. Returning later, he found that the sandbank on which the creature had been basking was thirty feet across, indicating that the creature must have been in excess of thirty-three feet in length.

Along the Lupar River, also in North Borneo, there once existed a similarly venerated reptile. Bujang Senang, or "Happy Bachelor," was the "king" of crocodiles. He was said to be twenty-five feet long, and a known man-eater who had accounted for many victims. The Iban tribe worshipped this veritable leviathan—a situation worthy of a story by Edgar Rice-Burroughs or Henry Rider-Haggard!

The late Abang Idris—chief of police for the town of Sir Amen—led a ten-year hunt for the huge reptile.

Like Moby-Dick, the Great White Whale chronicled by Herman Melville, Bujang Senang had outstanding markings: a large white blotch on his tail and a white back. And like Melville's cetacean, he proved very hard to kill. A Mr. Pit, resident of Sarawak, claimed to have been amongst a group of men who slew Bujang Senang on the third of June, 1990. Apparently, a huge crocodile grabbed and ate a woman along the Batang Lupar river, sixty miles east of Kuching. Her five-year-old child, who witnessed this, escaped. Mr. Pit and his gang solicited the help of a boto, or witch doctor, to help them draw the monster from the river. Once the creature appeared, the men attacked it with parangs (huge machetes similar to samurai swords) and axes. After a thirty-minute battle that hospitalised one of the men, they managed to hack through the ironwood-hard scales and kill the animal. Its flesh was then sold to some Chinese people for cooking. They reportedly found silver watches, coins, and human hair in the crocodile's gut.

On Mr. Pit's website (now defunct) where he recounted this adventure, there was a photograph of the men standing next to the animal's corpse. Unfortunately, there is no diagnostic white mark on the tail and the animal looked to be only seventeen feet long—far short of Bujang Senang's reported size.

The "Happy Bachelor" was finally killed in May 1992. He proved to be somewhat short of the twenty-five-foot mark. He was in fact "only" nineteen feet, three inches in length, but he did have the trademark white colouration. From the base of his neck to the base of his tail, his dorsal-scutes were worn and bleached. This gave an almost white appearance.

The great naturalist Charles Gould, who wrote the seminal cryptozoological book *Mythical Monsters* (some seventy years before the term cryptozoology was coined), was well aware of giant crocodiles. A friend of his, Mr. Dennys, a resident of Singapore, told him of a thirty-foot crocodile that haunted a tidal creek that ran through the city in the 1880s. Another colleague—Mr. Gregory, the Surveyor General of Queensland—informed him that Australia's northern rivers were home to crocodiles as long as a whale boat (twenty-eight feet).

Another giant, still alive at the time of writing, lives in the Bhitarkanika Wildlife Sanctuary in Orissa State in eastern India. It is over twenty-three feet in length, and three other animals in the same sanctuary have been reported as having achieved a length of over six metres (twenty feet).

Australia too still harbours giants. Malcolm Douglas, filmmaker and owner of the Broome Crocodile Park, encountered such a leviathan in a northern Australian river in 1987:

"We call him the hippo. He dwarfs everything else I've ever seen. Compared with him, the sixteen-foot-four-inch croc we did catch looked like a twelve-footer. Once we did have him alongside a net. The corks were three feet apart and his length covered eight corks along the net. Maybe a little more."

This would have made the "hippo" at least twenty-five feet long.

There is a long history of giant crocodiles in Australia. In 1860, a thirty-three-foot crocodile was said to have been shot on the banks of the Mossman River.

Another of the same size was shot by publican Jack O'Brian from the veranda of the Leichhardt Hotel on the banks of the Pioneer River. The beast was dubbed the Mackay Monster, after the town the river ran through. The creature was put on display at the hotel and the skin was kept there for some years, but subsequently vanished.

In 2000, a photograph of a huge crocodile with its jaws propped open was found by a council worker at a dump. The picture also has two human figures in it, a woman and man. The crocodile does indeed look huge.

Another thirty-three-foot specimen was shot in the 1880s on the Russell River near Cairns.

In 1960, a twenty-five-foot crocodile was shot on the Annie River in Princess Charlotte Bay, and another of the same size on the Staaten River.

Percy Trezise, pilot, artist, author, and expert on Aboriginal rock art, saw a giant crocodile whilst flying over the Cape York Peninsula with a group of tourists.

> *"I was flying low over an estuarine area, with the tide out, and I came around a corner at a hundred feet, and there was the crocodile of all crocodiles. It looked as if you couldn't reach across his back with spread arms. In my opinion he was more than thirty feet long."*

Back in the 1950s, he had seen another huge croc.

> *"In the late '50s I was flying for the Aerial Ambulance and up along the western coast, north of Weipa, I saw this big log with a fairly large crocodile nearby. When I circled the log, I saw its feet and realized that it, too, was a crocodile. I reckon it was twenty-four feet long."*

Lloyd Grigg saw a twenty-five-foot crocodile in a river running into Princess Charlotte Bay. It left a slide mark in the mud five feet six inches wide.

In 1929, Claude Le Roy used gelignite to blow up a twenty-five-foot crocodile, in a hole just below Hartley's Creek Crocodile Farm, north of Cairns.

Another twenty-five-foot monster terrorized the Staaten River for fifty years. Known as "the Wyabba monster," the local Aborigines believed that it could never die, as it was part of their "dreaming" or spiritual existence. Though shot at many times, the croc always shrugged off the bullets and carried on as normal. It was finally put paid to by hunter Peter Cole in the mid-1950s.

An even larger specimen persists at the time of writing. At over twenty-eight feet, it inhabits the Guider River swamps of northern Arnhem Land.

In former times, such a giant would have been more common. Hunting in the first half of the twentieth century reduced the number of "salties," but since their protection in the early 1970s, the species has made a spectacular comeback in Australia. Soon there may be more reports of giant crocs like these in protected areas.

The largest specimens of a truly mind-boggling size have been met with, not in rivers, but in the open sea. The Indo-Pacific is the most pelagic of all crocodiles, having been encountered hundreds of miles from land. Larger than the biggest predatory shark—the great white (*Carcharodon carcharias*), maximum twenty feet—it has nothing (except man) to fear.

One such sea-going encounter took place in the Gulf of Bengal in 1860. The crew and passengers of the ship *Nemesis* observed a giant crocodile at close range. One of the witnesses was the writer W. H. Marshall, who described it in his book *Four Years in Burma*:

> *As the Nemesis was proceeding onwards toward our destination our attention was directed to an alligator of enormous length, which was swimming along against the tide (here very strong), at a rate which was perfectly astonishing. I never beheld such a monster. It paused within a very short distance from us, its head and nearly half its body out of the water. I should think that it could not have been less than five and forty feet long measured*

from the head to the extremity of the tail, and I am confident that
it was travelling at a rate of at least thirty miles an hour.

It should be noted that this animal would have been an Indo-Pacific crocodile and not an alligator. Alligators only occur in China and North America.
However, early European colonials often used incorrect names for animals,
and these have stuck. In Australia, the Indo-Pacific crocodile is often referred
to as an alligator, and there is even an Alligator River. In Belize, jaguars are
called "tigers" and spider-monkeys "baboons." In Australia, monitor lizards
are mistakenly known as "goannas"—a corruption of "iguana," a purely New
World animal.

Another man who mistakenly referred to Indo-Pacific crocodiles as
"alligators" was British sea-captain and trader Alexander Hamilton. In the
East Indies in 1705, his men were disturbed by "alligators" whilst working on
a stage rigged alongside their ship. They fired musket-balls at the creatures
but could not penetrate their armour. One lucky shot finally struck a creature
in the eye and entered the brain. The following day they came across the
carcass on the shore. They measured it at twenty-seven and a half feet.

Another sighting of a giant Indo-Pacific croc comes from the Indonesian
island of Ternate. In February of 2002, the twenty-three-foot giant had
taken up residence in a river close to two villages. It had eaten four people.
The headless corpse of its latest victim, a teenaged boy, was found close to
the giant's lair.

More recently, in the massive wetlands known as Agusan Marsh in the
Philippines, a giant crocodile named Potol ("slasher") has made the news.
The local people have measured him against both their vessels and floating
houses and reckon that Potol is around thirty feet long. His is said to have
devoured many people, including a schoolgirl whose head he bit off after
flipping over her canoe. Such is the fear that the huge reptile engenders
that the lake people will move their floating villages if the monster appears
in the area.

It is fitting indeed that the other species of crocodile with a claim to having
attained monstrous proportions inhabits the dark continent of Africa. The
"Cradle of Man" has a deep hold on our subconscious. If giant reptiles are

to exist anywhere, it must be here. The heart of the African continent still remains an enigma.

Crocodylus niloticus (the Nile crocodile) is the world's second-largest known species of reptile and has long been known as a man-eater. Worshipped by the Egyptians as "Sebek" (the God of the Nile) and by innumerable sub-Saharan tribes, this is indeed a frightening animal.

Its prey includes lions, giraffe, buffalo, and even black rhino! As with the Indo-Pacific crocodile, it is not averse to adding humans to its menu.

The largest "official" specimen was shot in 1905 at Mwanza, a hundred km east of Emin Pasha Gulf by the Duke of Mecklenburg. It measured twenty-one feet in length. This monster, however, pales in comparison to some of the other crocodiles that have been reported by some naturalists and explorers.

The renowned wildlife photographer Cherry Kearton (1871–1940) and his friend James Barns observed a twenty-seven-foot crocodile basking on a sandbank in the Semliki River in Uganda. The size was estimated against other crocodiles and nearby objects. A photograph was published in one of Kearton's books, *In the Land of the Lion*, and apparently the crocodile in question dwarfs its companions. (I found a copy of this book in a secondhand book shop in York. The only photo of a crocodile in the book has no other crocodiles for comparison, so is of little use. My copy is a sixth edition, however, and may not contain the same picture as Kearton's original.)

A twenty-six-foot specimen was claimed by a Captain Riddick, who is alleged to have shot it at Lake Kioga in Uganda, and another of similar size was killed on the Mbaka River (in what is now Tanzania) in 1903. This was recorded by the experienced field naturalist Hans Besser. At first, he mistook the reptile for a huge canoe half-drawn out of the water. It was twenty-four feet long, but part of the tail was missing. (Perhaps it had been bitten off by an even bigger crocodile!) The body was three feet, six inches high, and was 14.72 feet in girth. The skull was 4.48 feet long.

In 1954, Guy de la Ruwiere saw a twenty-three-foot crocodile in the Maika marshes in the northeast Congo. The animal lifted its massive head out of the water several times. It caused a huge wave when it dived beneath the surface.

One must be careful when estimating size. My colleague Dr. Lars Thomas, of the University of Copenhagen, was told by some hunters that they had shot a thirty-five-foot crocodile in northern Australia. Dr. Thomas had the men show him the carcass and it turned out to be only eighteen feet long. The men had been quite sincere but had not brought a tape measure with them.

World-famous explorer Colonel John Blashford-Snell (since 2001, the Hon. Life President of the Centre for Fortean Zoology) heard tales of thirty-three-foot crocodiles in the Ethiopian Blue Nile Gorge, but personally saw none above twenty feet!

One man who is very adept at estimating size is Rupert Bunts. Mr. Bunts had been a soldier in Rhodesia (now Zimbabwe) in the early 1970s, and one of his jobs was to intercept terrorists from neighbouring Zambia. The easiest way to tell if a man was indeed a terrorist was by his boots, Zambian boots being different from Rhodesian ones. On one occasion, a suspect ran into the water in the southern end of Lake Kariba in an attempt to swim away from the patrols. The ill-fated fellow was seized and bitten in two by an immense crocodile. Mr. Bunts and his companions opened fire on the giant reptile with high-powered SLR rifles. These weapons can send a bullet through a brick wall at the range of a mile. Not even the armour-plate of such a monster could withstand this barrage. Once the titan lay still, they drew alongside in a boat. When they dragged it ashore and cut it open, the luckless victim's legs were retrieved. He was indeed a Zambian.

I asked Mr. Bunts how large the crocodile was. To my amazement, he told me it was between twenty-five and thirty feet long. Mr. Bunts was sure of this, as he was used to estimating distance and size as part of his job. Unfortunately, none of the men knew the zoological importance of the specimen, and no photos were taken, or samples kept.

More recently, a man-eating giant was uncovered by French environmentalist Patrice Faye. The twenty-three-foot creature was named "Gustave" by Faye and his team. His home is on the Burundi side of the Rusizi Delta. Gustave is estimated to be between eighty and a hundred years old, and may have eaten more humans than any other individual crocodile alive. In 2003–04, he is known to have eaten seventeen people. Locals say he has been dining on man-flesh for over thirty years, so his human victim-tally must be mind-boggling.

Patrice and his colleagues are attempting to capture Gustave alive and put him on display as a tourist attraction. They hope that this will boost the finances of the Rusizi Game Reserve that is currently being regenerated. He is already bringing in tourists, despite the wars raging in both Burundi and the Democratic Republic of the Congo. Twice before, the French team has attempted to capture Gustave and failed. The third attempt took place between May and December 2003, and was again a failure.

The largest reported crocodiles on the African continent hail from that last great African frontier, the Congo rain forest. They are known to the Lingala and other Congolese people as "Mahamba." This lord of the jungle is said to reach a shocking fifty feet in length!

In the late nineteenth century, Belgian explorer John Reinhardt Werner reported sightings of giant crocodiles that lend some weight to the terrifying folk tales of the native population.

Whilst travelling down the Congo on the *Aja*—a forty-two-foot steam launch—Werner stopped at a sandbank to shoot ducks. He shot one and pursued others over a low ridge when he saw "the biggest crocodile I have ever seen. Comparing him to the *Aja*, which lay in deep water some three hundred yards off, I reckoned him to be quite fifty feet long: whilst the centre of the saw-ridged back must have been some four feet off the ground where his belly rested."

Werner stupidly took another shot at the ducks (they had run out of meat on the ship) and alarmed the monster, which made off into the water. The creature was also witnessed by a native boy that Werner had with him.

Around three days later, Werner saw another vast specimen. The *Aja* had embedded itself in a sandbank when it was heaved up out of the water by something causing a commotion under the ship:

> *"I saw an enormous crocodile—longer I am certain than the Aja—rush across the bank and tumble into the deep water beyond. I never before saw such a large crocodile move so fast, and I had no time to get a shot at him. He must have heard us coming and was trying to make for the deep water on our side of the bank, when we ran into him and hammed him onto the sand. We struck*

him, moving at a rate of four miles per hour, but during the short time he was in view I could not see that he bore any marks of the collision!"

It would be as well now to pause and reflect on the dimensions of such a huge crocodile. A twenty-five-foot creature would be an awesome animal in the two-to-three-ton weight bracket. A fifty-foot animal would be of a colossal weight. When an animal doubles its size, its weight increases eightfold. This is because length, breadth, depth, and height have all been doubled. If we take the conservative estimate of two tons for the weight of a twenty-five-foot specimen, then a fifty-foot animal would weigh in the region of fifteen tons: three times the weight of an average elephant! If crocodiles of these dimensions do exist, then they are the largest macro-predators on the planet. Most of the great whales are plankton feeders, and even the toothed sperm whale feeds mainly on small fish and squid (the giant squid forms only 1 percent of its diet and weighs far less than the sperm whale in any case). Such a giant crocodilian would be surpassed only by the giant marine reptiles of the Mesozoic, and possibly the largest carnivorous dinosaurs. (Palaeontologist Gregory S. Paul postulates a maximum weight of twenty tons for the largest tyrannosaurs, and this seems to have been confirmed by a recently excavated specimen of this dinosaur.) If they do indeed exist, there is no animal on earth that could possibly withstand an attack from one of these giant saurians.

It is obvious that we are not dealing with whole races of gargantuan crocodiles, but rather a few massively large individuals. If they were prehistoric survivors of a giant race, then many more specimens would have turned up!

So what is it that causes certain crocodiles to become so large?

I believe that it is a combination of several factors.

Both Nile and Indo-Pacific crocodiles have large distributions. Within their range many subspecies can exist, and these may display large variations in size. A striking example is found along the Aswa River in Northern Uganda. Crocodiles here reach sexual maturity at a length between 4.9 and 5.9 feet, and never exceed seven feet. This is less than half the average size. It would seem that this is a strategy that has developed to avoid food shortages.

Other areas, such as Lake Malawi, the Congo, parts of Tanzania (such as the Grumati River), and the Semliki River in Uganda/Zaire, produce larger-than-average specimens.

Where populations of these larger-than-average animals have remained undisturbed, occasional freaks will be thrown up within the range of genetic variation that are much larger than the average. The average man is five foot nine inches tall, but many in the population exceed this. Most big cities have several seven-foot individuals, and the record human height is eight feet, 11.9 inches. A large population of "Big Crocodiles," most of whom would reach seventeen feet, could throw up some twenty-three- to twenty-six-foot specimens occasionally.

Diet is also a factor. It is believed that very big crocodilians are immensely old. It is thought that crocodilians grow roughly twelve inches a year until they achieve the length of ten feet, when the growth rate radically slows down. By this logic, to be immensely huge, a crocodile must have achieved a great age. Crocodiles can reach impressive ages.

A male Nile crocodile terrorized the Okavango Delta in Botswana. The man-eater was captured alive by an elephant hunter called Sir Henry in 1903, when already an adult. He was named Henry after his captor. He resides in the Crocworld Conservation Centre in Scottburgh, KwaZulu-Natal, and he weighs 1,100 pounds and measures close to sixteen feet in length. He has fathered at least 700 offspring. Henry now lives with a harem of six female crocodiles. He has been in captivity for 115 years, and must have been at least fifteen years old (and possibly far older) when he was caught. Henry is still living and is at least 130 years old.

Australia Zoo held a male freshwater crocodile (*Crocodylus johnsoni*) called Mr. Freshie who passed away in 2010 at the age of 140!

Protein intake seems to have more to do with large size than age. In the early 1970s, the Louisiana Department of Wildlife and Fisheries made some interesting discoveries relating to diet and growth rate in the American alligator (*Alligator mississippiensis*). Two groups of juveniles were reared on different diets. One was fed coypu—a large South American rodent (*Myocastor coypu*)—flesh, and the other, fish.

Nutritional analyses showed that coypu contained 14.9 percent crude protein, 2.1 percent crude fat, 0.1 percent crude fibre, and 45 percent moisture.

Fish, on the other hand, contained 9.9 percent protein, 4.0 percent fat, 1.0 percent fibre, and 60.6 percent moisture.

Specimens fed on coypu grew 20 percent larger than their fish-fed peers, over a period of three years. They were also more active and aggressive.

Food with more protein content causes accelerated growth. The Aswa crocodiles were tiny due to aestivation brought on by seasonal food shortages. In areas where protein-rich food is plentiful all year, the average size of the crocodile population was much greater. So if we think of a population of naturally big crocodiles feeding on protein-rich prey that occasionally produces a giant freak whose size is increased further still by its diet, then one can conceive of a truly vast animal.

Tropical seas and teeming rain forests would offer such an abundance of prey. It can be no coincidence that the largest reported crocodiles are seen in these very habitats.

Lack of disturbance may be a factor too. Most giant crocodiles are reported from remote areas where human interference has been minimal.

Monster Snakes

Constricting snakes (the boas and pythons) are the largest snakes alive, although not all reach excessive lengths (some are barely two feet long). Five species are known to exceed twenty feet in length. These are:

Python reticulatus	*the reticulated python at 33 feet*
Eunectes murinus	*the anaconda at 29 feet*
Moreli amethistina	*the amethystine python at 28 feet*
Python sabae	*the African rock python at 25 feet*

Python bivittatus	*the Burmese python at 26.5 feet*

Charles Gould, in his magnum opus *Mythical Monsters*, tells us of his belief that constricting snakes once grew far beyond their modern dimensions:

> *I fancy that at the present day the numbers, magnitude, and terrifying nature of serpents but feebly represents the power which they asserted in the early days of man's existence, or the terror which they then inspired.*

These snakes kill by suffocating, not crushing, victims in their muscular coils, and make good analogues for the "wyrm" type of dragon—even more so when one considers the amazing lengths reported for some specimens. These lengths far exceed the accepted maximums given above. Our search for giant snakes will take us all around the tropics, but we will start in the cradle of mankind: Africa.

A monster snake was photographed by the passenger of a Belgian military helicopter pilot in the Katanga region of Zaire in 1959. The reptile is pictured so clearly that even the scales on its hide are visible. The photographer was one Colonel Remy van Lierde, a decorated WWII pilot. During the war, he had shot down six enemy fighter planes and forty-four V1 flying bombs. He gained the rank of Squadron Leader in the RAF and, after the war, was made Deputy Chief of Staff to the Minister of Defence in 1954. He was working as a charter pilot at the time. His chopper crossed a hollow in a jungle clearing, and van Lierde saw a vast snake emerging from a hole.

The huge serpent was dark green with a lighter-coloured underside. Van Lierde, who was adept at estimating size, put the creature at fifty feet long. He made several passes, allowing his passenger to shoot a picture. The snake reared up fully ten feet, as if to strike at the helicopter. Van Lierde estimated the head of the snake to be three feet long by two feet wide, and compared it to the head of a gigantic horse.

An African rock python of thirty-two feet was supposedly shot near Bingerville, in the Ivory Coast. Though unconfirmed, it beats the official record by seven feet. The Ivory Coast seems a haven for large pythons, because another of the same species, twenty-four feet long, was killed in Adiopodume.

In August 2000, an oil worker from Egbema-Ogba, Nigeria, was swallowed by a twenty-five-foot rock python. George Otoh, thirty-three, was relieving himself in bushes when the massive snake attacked. His body was later discovered inside the reptile.

One remarkable woman who had experience with giant snakes in Africa was Mary Kingsley, niece of author Charles Kingsley. She had led a sheltered life until the age of thirty, then suddenly decided that she wanted to explore Africa, and study its religions and superstitions. In a time when women were meant to stay at home, she explored the then-truly-wild areas of West and Central Africa, collecting specimens for the British Museum. She recorded her remarkable adventures in a book, *Travels in West Africa*. Therein she tells of outsized specimens of several species:

> *The largest crocodile I ever measured was twenty-two feet three inches, the largest gorilla five feet seven inches. I am assured by the missionaries at Calabar that there was a python brought into Creek Town in the Rev. Mr Goldie's time that extended the whole length of the Creek Town mission-house veranda and to spare. The python must have been over forty feet. I have not a shadow of doubt it was. Stay-at-home people will always discredit great measurements, but experienced bushmen do not, and after all, if it amuses stay-at-homes to do so, by all means let them; they will have dull lives of it and it don't hurt you, for you know how exceedingly difficult it is to preserve really big things to bring home, and how, half the time, they fall into the hands of people who would not bother their heads to preserve them in a rotting climate like West Africa. The largest python skin I ever measured was a damaged one, which was twenty-six feet.*

Modern-day experts would do well to take a leaf from Miss Kingsley's book. Time and again, armchair zoologists will proclaim that this or that cannot exist, without ever leaving the ivory towers of their lecture halls.

Asia, too, has its tales of giant serpents. In fact, it is here in Asia, on the island of Celebes (now Sulawesi), that the official longest snake in the world was captured: a thirty-three-foot reticulated python taken in 1912. Of course, larger specimens have been reported. Eighteenth-century explorer Francis Legaut claimed to have encountered one fifty feet long on Java. A brute of similar size was reported in the *North China Daily News* of November 10, 1880. The story tells of a Western hunter (whose name is never revealed), who came across a remote hut in the dense jungle between Buddoh and Sirangoon on the Malay peninsula. Upon the roof was the skin of a gigantic python. Inquiring as to its origin with the hut's owner, he was told this story:

The Malay was awakened one night by his wife's screams. Investigating, he found to his horror an immense snake that had drawn the poor woman's whole arm into its maw and was in the processes of swallowing her. The plucky fellow seized two bags and stuffed them into the corners of the giant reptile's mouth thus forcing them to open wider. The snake released the woman and turned upon the man whipping its coils about him. Fortunately for the Malay his arms were free, and he grabbed his parang and hacked at the vasty serpent. The snake unwound and slithered through an opening beneath the hut. Both the man and the hut were covered in blood.

Come morning he followed the python's trail to a patch of plantain palms. In its death throes the beast had smashed the trees and uprooted them. In the midst of the destruction lay the offending creature, dead. He had been offered sixty dollars from some Chinese who had travelled long distances to buy pieces of the monster's flesh due to its medicinal properties. They also offered him six dollars for the skin, this he kept however, as a trophy of the ordeal. The skin was between seven and eight fathoms (fifty to fifty-six feet) long.

Unfortunately, skins can be stretched when removed from the corpse of a snake, giving an unnaturally long appearance. Even so, this snake—if the estimates were right—must have been a colossal animal. So big in fact that one doubts that it could have been killed anything like as easily as it is claimed.

An even more dramatic story is recounted in the Victorian natural history tome *Pictorial Museum of Animated Nature*:

> *The captain of a country ship, while passing the Sunderbunds, sent a boat into one of the creeks to obtain some fresh fruits, which are cultivated by the few miserable inhabitants of this inhospitable region. Having reached the shore, the crew moored the boat under a bank, and left one of their party to take care of her.*

> *During their absence, the lascar who remained in charge of the boat, overcome by the heat, lay down under the seats and fell asleep. Whilst he was in this happy state of unconsciousness an enormous boa (python) emerged from the jungle, reached the boat and had already coiled its huge body round the sleeper, and was in the very act of crushing him to death, when his companions fortunately returned at this auspicious moment, and attacking the monster severed a portion of its tail, which so disabled it that it no longer retained the power of doing any mischief. The snake was then easily dispatched, and was found to measure, as stated sixty-two feet and some inches in length.*

If this event actually occurred, then the creature would have been an outsized reticulated python, not a boa. Once again, the ease of its death raises suspicion.

The official record for the reticulated python stands at thirty-three feet for a specimen killed near a mining camp in the Celebes. Animal dealer Henry Trefflich is said to have obtained a thirty-two-foot specimen from an unspecified source, but this has never been proven. Other claimed giants include a thirty-three-footer killed on Java and a thirty-foot python killed near Penang, Malaysia, in 1844.

One verified titan was "Colossus"—a 28.5-foot reticulated python held at Pittsburgh's Highland Zoo. At the time of her death (from old age) in 1966, she was estimated to weigh three hundred pounds.

The now defunct Knaresborough Zoo in Yorkshire once held a twenty-seven-foot, four-inch reticulated python called Cassius who weighed in at 220 pounds.

Australia, unsurprisingly, has also produced giant snake stories. Charles Gould was told by his acquaintance, G. R. Moffat, that the Aborigines on the Lower Murray River, between Swan Hill and the Darling junction, knew of a giant black serpent that lived in the Mallee scrub. It was forty feet long, of huge girth, but very swift. Fortunately, it produced a vile stench that warned of its approach. A white man—the son of Mr. Peter Beveridge of Swan Hill station—had seen the beast. This was around 1857.

Mr. Henry Liddell, a resident on the Darling River, was told identical stories by stock-riders and ration-carriers. The ebony monsters were considered not uncommon between Wentworth and Pooncaria in the 1870s.

The mother of all down-under giants was reported back in 1822, by two men in front of a bench of magistrates in Liverpool near Sydney. The men told them that, just four kilometres outside of town, they had come across a snake forty-seven feet long and three times as thick as a human. Thinking it was dead, one of the pair unwisely threw a rock at it. To their horror, it was very much still alive, and rose five feet off the ground. The magistrates seemed to believe them, as a posse of armed townsfolk ventured to the location of the encounter but found only a large track bearing the impression of scales.

The longest snake actually measured in Australia was a twenty-eight-foot amethystine python killed at Greenhill near Cairns in 1948.

It is in South America that we meet with the most numerous reports of outsized ophidians. This is unsurprising, as the neo-tropics is the lair of the giant snake *El Supremo*, the anaconda. In terms of bulk, this snake is by far the largest in the world. Its girth is far greater than that of the reticulated python. Ever since the white man first ventured tentatively into the green hell, he has brought back tales that are the very stuff of nightmares—snakes whose size defies belief.

The earliest man to return with such bone-chilling yarns was one Charles Waterton (1782–1865), better known as Squire Waterton, a great British eccentric and adventurer. A Yorkshireman from a wealthy Roman Catholic family, the Squire insisted on sleeping on bare boards with a block of wood as his pillow. Almost unique in his age, he was a teetotaler and violently opposed to hunting for sport. He was a passionate naturalist and collector of animals and, with true intrepid Yorkshire spirit, he made four expeditions to South America between 1812 and 1824, travelling in Brazil, Venezuela, and Guiana.

In typical Waterton style, he exposed as much of his skin as he could in the jungle at night, hoping to be bitten by a vampire bat. He was most disappointed when he was not bitten—but one of his companions was. The ungrateful man ran and hid in a latrine. Waterton's books are full of such shenanigans, and it is obvious he enjoyed himself immensely. The Squire lived to the ripe old age of eighty-three, a miracle when one reads of some of the risks he took!

Of the anaconda, he writes:

> *The camoudi snake (as it was called in British Guiana) has been killed from thirty to forty feet long; though not venomous, his size renders him destructive to the passing animals. The Spaniards in the Oroonoque positively affirm that he grows to the length of seventy or eighty feet and that he will destroy the strongest and largest bull. His name seems to confirm this; there he is called "matatoro" which means literally "bull killer." Thus, he must be ranked among the deadly snakes, for it comes to the same thing in the end whether the victim dies by poison from the fangs, which corrupts his blood and makes it stink horribly, or whether his body be crushed to a mummy and swallowed by this hideous beast.*

Of course, the anaconda kills by constriction, not venom.

A missionary, Father de Vernazza, wrote in the nineteenth century what is surely the most fatuous description of the anaconda:

The sight alone of this monster confounds, intimidates and infuses respect into the heart of the boldest man. He never seeks or follows the victim upon whom he feeds, but so great is the force of his inspiration, that he draws in with his breath whatever quadruped or bird may pass him within twenty to fifty yards of distance, according to its size. That which I killed from my canoe upon the Pastaza (with five shots from a fowling piece) had two yards of thickness and fifteen yards of lengths; but the Indians have assured me there are animals of this kind here of three or four yards in diameter, and from thirty to forty long. These swallow entire hogs, stags, tigers, and men, with the greatest facility.

The good father was confusing diameter with circumference methinks, else his snakes would be extremely stout. Alternatively, it may have shot one that had just eaten a large prey item, such as a tapir. The super snake suction he speaks of is total fantasy, but the supposed mystic effect of anaconda breath is a stubborn myth—as we shall see in a moment. The "tiger" referred to is actually the jaguar (*Panthera onca*).

In 1944, another large specimen was encountered in Colombia by a team of prospecting geologists led by Roberto Lamon. The men shot the snake and measured it at thirty-seven feet, six inches. The group left the creature to eat their lunch, intending to come back and photograph their trophy and skin it. Upon their return, they were amazed to find it gone. The bullets had merely stunned the animal, which had recovered and absconded in their absence.

Fredrico Medem, a Colombian herpetologist, saw an anaconda that he estimated to be between thirty and forty feet, and obtained a report of another thirty-four feet long.

General Candido Mariano de Silva Rondon, who lent his name to the Rondonia area of Brazil, saw a specimen killed by Indians that was some thirty-eight feet long. There are several records of snakes in this size bracket that cannot easily be dismissed, as some have involved reputable scientists. A thirty-four-foot anaconda was shot by Vincent Roth, director of the National Museum, in British Guiana (now Guyana). Mr. R. Mole—a naturalist who made many important contributions to the knowledge of the wildlife of

Trinidad—reported a thirty-three-foot example there in 1924. Dr. F. Medem of the Columbia University saw a thirty-three foot, eight-inch snake killed on the Guaviare River.

In 1909, war was on the verge of exploding in South America. A "rubber rush" to rival the gold rushes of the Old West was happening, and a dispute was occurring in the Rio Abuna rubber plantations on the western borders of Brazil. Peru and Bolivia also meet at this point, and a bitter wrangle between the three countries over the valuable resource was growing to dangerous levels. Into this drama, the Royal Geographical Society sent a mediator to defuse the situation. Major Percy Fawcett, a thirty-nine-year-old artillery officer, was to make the first intensive study of the area.

It was whilst engaged in this task that he initially heard of giant snakes. The manager of a remote hamlet called Yorongas told him that he had killed a fifty-eight-foot anaconda in the lower Amazon. Fawcett disregarded the story at first, but subsequently claimed to have shot an even bigger specimen.

Several months after the conversation at Yorongas, he was on the Rio Abuna, upstream from its junction with the Rio Rapirrao, when:

> ...almost under the bow of the igarite there appeared a triangular head and several feet of undulating body. It was a giant anaconda. I sprang for my rifle as the creature began to make its way up the bank, and hardly waiting to aim smashed a .44 soft-nosed bullet into its spine ten feet below the wicked head. At once there was a flurry of foam, and several heavy thumps against the boat's keel, shaking us as though we had run on a snag. With great difficulty I persuaded the Indian crew to turn in shoreward. They were so frightened that the whites showed all round their popping eyes, and in the moment of firing I had heard their terrified voices begging me not to shoot lest the monster destroy the boat and kill everyone on board, for not only do these creatures attack boats when injured, but there is also a great danger from their mates.
>
> We stepped ashore and approached the reptile with caution. It was out of action, but shivers ran up and down the body like puffs

of wind on a mountain tarn. As far as it was possible to measure, a length of forty-five feet lay out of the water, and seventeen feet in it, making a total length of sixty-two feet. Its body was not thick for such a colossal length—not more than twelve inches in diameter—but it had probably been long without food. I tried to cut a piece out of the skin, but the beast was by no means dead and its sudden upheavals rather scared us. A penetrating foetid odour emanated from the snake, probably its breath, which is believed to have a stupefying effect, first attracting then paralysing its prey. Everything about this snake was repulsive.

Such large specimens as this may not be common, but trails in the swamps reach a width of six feet and support the statements of Indians and rubber pickers that the anaconda sometimes reaches an incredible size, altogether dwarfing that shot by me. The Brazilian Boundary Commission told me of one killed in the Rio Paraguay exceeding eighty feet in length.

This is the most celebrated and oft-repeated encounter with a giant anaconda, but it is also one of the most questionable.

Firstly, the width given for this snake is absurdly small. The anaconda is a massively-built snake. A specimen half this length would have a width twice as wide or more. Fawcett's snake would have had to be an emaciated near-skeleton!

Secondly, his assertion that "there is a great danger from their mates," implies that anacondas mate for life and their partners will seek revenge for the killing of a mate. This is nonsense—no snakes are life-maters, and anacondas breed in huge "mating balls." These consist of dozens of males competing to mate with one larger female.

Finally, no snakes have "stupefying breath." The breath of a giant ana-conda may well be foul, but it possesses none of these attributes.

For these reasons, I am inclined to reject Fawcett's story as a traveller's tale. The man himself disappeared several years later whilst looking for a lost city in the jungle.

There are other accounts, however, which are not so easily dismissed, and the anaconda has one huge advantage over the python that may well allow it to attain a greater size. All pythons are oviparous—that is, they lay eggs. This must be done on land. Anacondas are ovo-viviparous—they retain the eggs inside their bodies until the young hatch, then give birth to them live. This means they do not have to leave the water; their final link with the land is broken. Living in water almost all of the time means anacondas are buoyed up; they do not have to support their own body weight on land very often, and hence can grow to a very large size.

The Marquis de Wavrin was another explorer of South America and was active in the years before the Second World War. He told the great Belgian cryptozoologist Bernard Heuvelmans that he had seen anacondas over thirty feet long, and that the natives told of far larger ones. He once shot a twenty-six-foot individual that had been coiled around a branch. When he expressed a desire to retrieve the cadaver, his canoe-men told him that it was a waste of powder to shoot such a small snake and a waste of time picking it up.

They went on to say, "On the Rio Guaviare, during floods, chiefly in certain lagoons in the neighbourhood, and even near the confluence of this stream, we often see snakes that are more than double the size of the one you have just shot. They are often thicker than our canoe."

F. W. Up de Graff—an explorer of seven years' experience—spotted a giant anaconda as it lay in shallow water under his canoe. He said:

> It measured fifty feet for certain, and probably nearer sixty. I know this from the position in which it lay. Our canoe was a twenty-four-footer; the snake's head was ten or twelve feet beyond the bow; its tail a good four feet beyond the stern; its body was looped into a huge S, whose length was the length of our dugout and whose breadth was a good five feet.

When witnesses are cross-examined face-to-face by a renowned zoologist, we have to give them a little credence. One of the witnesses of the next case was interviewed over several days by no less an authority than Heuvelmans himself.

It was in 1947, when many wild and sometimes fierce tribes were still commonplace in the South America. A particularly warlike group were the Chavantes, who had recently killed a number of Brazilian officials. Francisco Meirelles of the Service for the Protection of the Indians organized an expedition to try to establish peaceful relations with this tribe. The five-month endeavour included in its twenty-man lineup Serge Bonacase—a French painter whom Heuvelmans later interviewed.

By the second month, the company had reached a large island between the two branches of the Araguaia River and made base camp there. The men spent several days in preparation for the big push into the wilderness (or the sertao, as the "green hell" was known). They made long reconnaissance and hunting trips away from the island. On one such trip, eight of them were hunting capybaras in a swamp between the Rio Manso (charmingly known as the Rio das Mortes, "The River of Death," as the Chavantes butchered anyone who dared to cross it) and the Rio Cristalino. The Chavantes did not put in an appearance, but the group encountered something far more frightening:

The guide pointed out an anaconda on a rise in the ground half hidden among the grass. We approached to within twenty metres of it and fired our rifles at it several times. It tried to make off, all in convolutions, but we caught up with it after twenty or thirty metres and finished it off. Only then did we realise how enormous it was; when we walked the along the whole length of its body it seemed as if it would never end. What struck me most was its enormous head.

As we had no measuring instruments, one of us took a piece of string and held it between the ends of the fingers of one hand and the other shoulder to mark of a length of one metre. Actually, it could have been a little less. We measured the snake several times with this piece and always made it twenty-four or twenty-five times as long as the string. The reptile must therefore have been nearly twenty-three metres long.

Unfortunately, none of the men were zoologists, and none realized the importance of the find. Bonacase himself had heard so many stories of giant anacondas he believed them to be commonplace. The carcass, and even the skin, would have weighed the men down too much for them to have brought it back. So, sadly, this invaluable specimen was left to the jungle scavengers. (This seems to be the bane of cryptozoologists. Specimens always seem to fall into the hands of those who do not know their importance, hence seldom find their way to civilization.)

The late 1950s brought perhaps the most dramatic encounter with an anaconda. The political climate, with the resurgence of communism in Latin America, was such that the US government placed CIA agents in sensitive areas. One agent, called "Lee," was told by a cattle rancher of a giant snake lairing in a cave in Bolivia. The reptile was said to be over thirty-three feet long. It was said to have eaten ten Indians and many cattle over the years. Every three months or so, the serpent emerged, seized a steer, dragged it into the river, killed it, and ate it. Then it would return to its cave.

The rancher wanted Lee to capture the animal and take it to a zoo, as it was "probably the largest snake in the world." The problem was discussed at the embassy many times, until someone came up with an audacious plot to catch it. The plan was to flush the monster from its lair with tear gas whilst a long sack (complete with zip fasteners) was held over the cave's mouth. There would be two "zip-men"—one at each end of the sack—to hasten the operation. For added security, Lee carried (ironically) a .357 Python pistol.

It was just as well Lee was "packing heat," as things did go spectacularly wrong. The tear gas was shot into the cave, and the anaconda, thrashing madly, shot out of the cave and into the sack. Once its entire length was inside, both ends were zipped up. The agents had not reckoned with the snake's vast strength, however. Its violent writhing split the sack end to end, and the brute was free.

The livid animal came rushing at Lee, who whipped out his pistol and managed to put a bullet in its head. The snake threw itself into a huge loop, smashing into a small hardwood tree about as big as a telephone pole. The tree was shattered like matchwood, and the snake fell back into the jungle. Lee pumped another two bullets into its head. When it had expired, they measured it. Its length proved to be thirty-four feet, three inches. Lee skinned the snake and took the hide back to the United States, where he kept it in his garage. Its current whereabouts are unknown. As noted earlier, this size would seem very small for a snake which was able to swallow such large livestock.

Lee's colleague David Atlee Phillips understandably doubted his friend's outlandish story. Some time later, he was attending a party in Washington, and mentioned the saga to Darwin Bell, then Deputy Assistant Secretary for International Labor Affairs. Bell claimed not only to have known Lee, but to have taken part in the capture attempt.

"I was the tail zipper man," he told an amazed Phillips.

More recently, a giant anaconda was reported near Sao Paulo, Brazil. Farmer/hunter Joao Menezes was fishing with his three-year-old son Daniel and turned his back on the boy to store some fish in a wooden shack. Suddenly his son's screams rent the air, and the horrified Menezes turned to see that a forty-five-foot anaconda had risen from the waters and sized his boy by the neck. He tried in vain to pry the snake's jaws apart, then ran home for his rifle. By the time he got back, however, the boy had been crushed, and was in the process of being swallowed.

More recently still, Colonel John Blashford-Snell was told a most intriguing story whilst travelling across the Andes by river from Bolivia to Bunenos. It seems that a thirteen-metre (forty-three foot) anaconda was captured by a farmer after it had eaten a cow. He apparently incited it with a pig on a rope. Subsequently he tried to sell his story, unsuccessfully, to the press. The creature is now said to be residing in a pond on a farm in northwest Brazil. This occurred in late 1999.

Reptiles keep growing throughout their lives, although the rate does slow down as they age. We have seen that some reptiles, such as crocodiles, can reach advanced ages. Giant tortoises can live to spectacular ages as well. An Aldabran giant tortoise (*Aldabrachelys gigantea hololissa*) living on the island of Saint Helena is 186 years old. Another of the same species that was kept at Calcutta Zoo was thought to be 255 at the time of its death. If, under unusual conditions, constricting snakes could reach great ages, they too could reach great sizes, and maybe they do, in the deep jungles of the tropics.

CHAPTER THREE

My Own Adventures in Monster Hunting

"People make mistakes in life through believing too much, but they have a damned dull time if they believe too little."

—James Hilton, *Lost Horizon*

The creature is about the size of a human baby, but there the resemblance ends. It has a brown, clay-like skin, stumpy arms and legs, a fanged mouth, a beard, and weird staring eyes. It lies on the back seat of a car and, unseen by the driver, it jerks into hideous life and begins to creep toward him.

This is my first memory of any kind. It was not a bad dream, but a 1971 episode of *Doctor Who*, entitled "Terror of the Autons." I was only one at the time of transmission. Maybe it was a later transmission, I don't know, but it is my first memory of any kind. I owe my career to *Doctor Who*. Starting in 1963, it is the longest-running and most successful science fiction show in history, but it had as much to do with horror as it did science fiction. Peaking in the 1970s with Jon Pertwee and Tom Baker, *Doctor Who* features giant green maggots crawling out of slag heaps in Wales, super-intelligent humanoid dinosaurs, killer dolls possessed by the disembodied minds of squid-like aliens, giant rats in Victorian sewers gnawing people's legs off, a homicidal cyborg ventriloquist's dummy with the brain of a pig, and robot mummies, to name but a few. Forget about what tries to pass itself off as *Doctor Who* today. It is now more interested in ticking boxes and pandering to political correctness. But back in the '70s it meant business, and its business was to frighten.

From the earliest age, I suffered from that wonderful malady called "wanderlust." Growing up with thrilling and exotic images of distant lands and strange creatures from magical series like *The World About Us* and *Life on Earth*, I would always be dissatisfied with suburban, industrial England.

I became obsessed with monsters and read all I could about them. Upon leaving school, I became a zookeeper specializing in reptiles. After growing tired of the incompetence and spitefulness of the old women who ran the zoo,

I left to study zoology at Leeds University. During the summer holidays, I was out searching Bodmin Moor for the Beast of Bodmin Moor, one of the British big cats. During a break, I visited the now sadly defunct Potter's Museum of Curiosities at the Jamaica Inn. The museum was a heterogeneous collection of strange exhibits. You could see stuffed kittens getting married next to the head of a man-eating crocodile from India, next to a model of a church made from feathers! It was here I picked a copy of a magazine called *Animals & Men* (which wasn't as rude as it sounds). It was the journal of an organization called the Centre for Fortean Zoology, based in Exeter, Devon. The CFZ was a professional organization that searched for unknown animals.

The magazine was very good, and I subscribed to it. I began a correspondence with Jon Downes, the director, and began to write articles for the magazine. After a couple of visits down to the Centre, Jon invited me to join the CFZ as its zoological director.

Since then, I have travelled the globe in search of all kinds of strange creatures and have had equally strange adventures. In this chapter, I will share some of them with you. Space dictates that I cannot detail all of my trips, so I will relate some of the most interesting here.

The Orang-Pendek

The cryptid I have been most linked with is the upright-walking Sumatran ape called the orang-pendek.

I first ventured into the rain forests of Sumatra in June 2003. The small expedition team consisted of Dr. Chris Clark, Jon Hare, and me. We contacted Debbie Martyr, a travel writer turned conservationist. Debbie, who now resided in West Sumatra, had become the head of the Indonesian tiger conservation group. She has seen the orang-pendek on more than one occasion. Debbie was very helpful, suggesting where to look and which guides to employ.

Whilst in a hotel in Padang, the unattractive capital of West Sumatra, we met a man in his fifties called Stephano, who claimed to have seen the orang-pendek. He told us that in 1971, he had accompanied an Australian explorer called John Thompson into the jungles of Kerinci Seblat National Park. He had seen small human-like primates with yellow hair. In order to stop Thompson shooting them, he told the Australian that a curse would descend on anyone who killed one of the creatures.

Sadly, before we could question him more, our transport arrived to convey us south to Sungai Penuh, where Debbie Martyr lived.

The next day we met Debbie Martyr. Debbie is a charming lady who reminded me a lot of the chimp conservationist Jane Goodall. A former journalist, Debbie first came to Sumatra as a travel writer in 1993. She had heard tell of orang-pendek and assumed it was a legend, no more than a bit of local colour. Later, a guide was telling her of the animals he had seen in the jungle. He said he had seen rhino, sun bear, tiger, elephant, and orang-pendek! About six weeks later, Debbie saw the animal.

Debbie told us that the most recent sighting, about three months previously, had taken place in the jungle surrounding Gunung Tujuh, or the lake of seven peaks, a large volcanic lake in the park. She photocopied several maps for us and spoke of a lost valley. Despite its being shown on the map, Debbie told us no one had ever been there. It looked like a couple of days' hike from the lake. The contours showed a wickedly steep-sided canyon.

We all felt that it would be exciting to look for the valley. She had also arranged guides. Sahar was a small bespectacled man in his early thirties, who, if dressed in a suit and tie, could pass for an accountant. The others were his brother John and an older man called Anhur.

We travelled down to Sahar's village, Polempek, and next day began the expedition proper.

Fully stocked, the six of us set out into the foothills of Gunung Tujuh. The foothills were fine, but as the gradient grew more acute, I began to suffer. Lake Gunung Tujuh is 6,578 feet high. Much of the way, the path is at something like seventy-five degrees. Imagine a gargantuan winding staircase. The stairs are made of moss-slick tree roots jutting out at differing angles. Like the labour of Sisyphus in Greek mythology, the climb seemed never-ending. I collapsed with exhaustion, staggered on, collapsed again, and vomited from over-exertion.

Finally, I made the summit. The land falls away dramatically to the four-kilometre lake. Lake Gunung Tujuh is a strange unearthly turquoise in colour. It lies in the bowl of an extinct (or maybe just dormant) volcano. Geo-thermal in nature, its waters are warm. The lake's waters are biologically impoverished. Only one species of small fish and one species of freshwater crab live in the lake. Despite this, the waters support several fishermen who sometimes come up from the village.

Using fishermen's canoes so decrepit they could have come off the ark, we crossed the lake. Whilst the others set up camp, Sahar led Jon, Chris, and me out into the jungle. Sahar's skill as a guide was astounding. The slightest bent twig or misplaced leaf catches his eye. Things that you or I would walk straight past tell him the secrets of the jungle. He pointed out the trail of a Malayan tapir (*Tapirus indicus*) through the bushes. The bulky animal had hardly disturbed the greenery. Later we found its three-toed footprints.

We came upon a possible orang-pendek footprint. Sadly, it had been damaged by rain. I measured it, but it was too damp for casting. It was narrower at the heel than at the front and pressed about half an inch into the ground. Further along the trail, we came across seven prints crossing a large muddy puddle. Similar in size and shape to the earlier print, they too had suffered damage. The gait was definably that of a biped. A fallen log crossed the puddle and, as Sahar pointed out, a human would have crossed by the log as opposed to walking through the mud.

A little further on, Sahar pointed out some damaged plants. Known as pahur, the pith inside the stem is a favourite food of orang-pendek. A number of the plants seemed to have been dexterously peeled apart and the pith eaten. A flattened area of moss on a nearby tree stump may have been where the creature sat whilst eating. We hid and waited in silence, but apart from the calls of birds and insects, nothing disturbed the stillness of the jungle.

Sahar told us that in 2000 he had heard the cry of orang-pendek. He demonstrated...

"UHUUUUUUUUR-UR-UR..." A weird, drawn-out moan, followed by two grunts. Quite unlike any animal vocalization I know.

In the 1980s, Sahar's father and a friend had been cutting logs to build a house close to where the village of Polompek now stands. The area has long since been deforested. Both men saw a bipedal ape lifting up cut logs and throwing them about. It was covered in blackish-brown hair and was about five feet tall. The hair on the creature's spine was darker. Its legs were short, and its powerful arms were long. The face was broad and was black in colour, with some pink markings. Both men fled.

Each day Sahar took us out on jungle trails. We collected a number of hairs but did not come across any more footprints. Thus we spent the days of our first expedition. We decided to leave the "lost valley" until next year.

As we climbed down again, we saw more wildlife in a single afternoon than in the whole of our stay at the lake: mitred langurs (*Presbytis mela-lophus mitrata*); a banded linsang *Prionodon linsang*, a normally nocturnal member of the civet family; a small toothed palm civet (*Arctogalidia trivir-gata*); and a pair of horse-tailed squirrels (*Sundasciurus hippurus*). We also found the droppings of a golden cat (*Catopuma temminckii*).

Back in Sungia Penuh, I interviewed Debbie about her sightings.

Me: *Could you please tell me how you first heard about and got interested in orang-pendek?*

Debbie: *I was travelling in Sumatra as a journalist in 1989. I was climbing Mount Kerinci and heard of a legendary animal that I thought would add a bit of colour to the travel piece I did. Then I started meeting people who claimed to have seen something that didn't appear to exist. At that stage I didn't believe or not believe; I was trained as a journalist, which is a respectable profession, so I took a look into it.*

Me: *Can you tell me about the first time you actually saw orang-pendek?*

Debbie: *I saw it in the middle of September; I had been out here four months. At that time, I was 90 percent certain that there was something here, that it was not just traditional stories. I thought it would be an orang-utan and that it would move like an orang-utan, not bipedal like a man. I had my own preconception of what the animal would look like if I did see it. What was the real shocker was that I had been throwing away reports on the animal on the basis of colour that didn't fit into what I thought the animal would look like. When I saw it, I saw an animal that didn't look like anything in any of the books I had read, films I had seen, or zoos I had seen. It did indeed walk rather like a person and that was a shock.*

Me: *What did it actually look like?*

Debbie: *A relatively small, immensely strong, non-human primate. But it was very gracile, that was the odd thing. So, if you looked at the animal you might say that it resembled a siamang or an agile gibbon on steroids! It doesn't look like an orang-utan. Their proportions are very different. It is built like a boxer, with immense upper body strength. But why an animal with immense upper body strength should be lumbering around on the ground I don't know. It makes no sense at all.*

It was a gorgeous colour, moving bipedally and trying to avoid being seen. I knew there was something in the vicinity because the action of birds and primates in the area meant that there was obviously something moving around. So, I sent a guide around as far as I could to where the disturbance was. Whatever was concealed in the undergrowth would try to avoid my guide and move away in front of him. I was concealed looking down over a small shallow valley. We didn't know what we were going to see. It could have been a bear, it could have been a tiger, it could have been a golden cat, or anything. Instead, from totally the wrong direction, a bipedal, non-human primate walked down the path ahead. It was concentrating so hard on avoiding my guide it didn't look toward me. I had a camera in my hand at the time, but I dropped it, I was so shocked. It was something so new my mental synapses froze up for a minute, trying to identify something I hadn't seen before.

Me: *You have seen it a couple of times since. Could you tell me about those sightings?*

Debbie: *I saw it again about three weeks later. Again, it was on Mount Tuju and again, I had a camera in my hand; again I froze because I didn't know what I was seeing. It had frozen on the trail because it had heard us coming. All I could see was that something across the valley had changed. I looked through a pair of binoculars. Something didn't look quite right in the landscape. By the time I trained on the area the animal, whatever it was, had gone.*

Those were the only times I could have got a photo of it. I have seen it since but fleetingly. Once you have seen an animal you can recognise it. If you have seen a rhino, you can recognise a bit of a rhino.

Debbie also showed us a cast of an orang-pendek footprint taken a few years previously in the jungle surrounding the lake. It was about eight inches long and did not resemble a yeti or sasquatch footprint. It was much less human-looking. It had four longish toes at the front, and the big toe was placed further back along the side of the foot. The toes all looked more prehensile than a human's, but less so than any known ape's.

Debbie also believes that early Dutch explorers may have collected orang-pendek specimens without knowing what they were. Bones and skin from this cryptid may be languishing in the basements of Dutch museums mislabelled as orang-utan!

We moved on to check out another sighting area.

Later, after a backbreaking trek, we reached a remote village called Sungi-Khuning. We stopped in a large (by village standards) house in the centre of the village. I was unsure whether it was a guest house, village hall, or just some hospitable soul's home. Sahar said that a man who had recently seen orang-pendek lived in the village and would come to talk with us. That night about twenty-three people crowded into the house, but the witness was not among them. Apparently, the man was a poacher who had set a snare for deer. Upon checking it, he found he had caught an orang-pendek. The powerfully built, five-foot-tall ape was struggling with the snare. The poacher tried to jab the orang-pendek with his spear, but the beast smashed it to matchwood and screamed at him. At this point he fainted. When he awoke the creature had pulled itself free and was walking off into the jungle. It was hair-covered, with long, powerful arms, and walked erect. The man himself, however, was not in the village at the time. Excursions into the surrounding jungle provided no further clues.

Once we were back in England, I sent hair samples off to my old friend Dr. Lars Thomas of Copenhagen University for analysis. The team at Copenhagen, headed by Dr. Tom Gilbert, specialize in retrieving DNA from old or damaged specimens. The samples all turned out to be from tapir or golden cat. However, I was convinced that there was an unknown species of ape in West Sumatra and resolved to study the area again.

Several weeks later Debbie emailed me to say that a honey-coloured orang-pendek had been recently reported from Renah Permatk. It had supposedly killed three dogs. The locals set out to catch it and Gunung Tujuh was crawling with people armed with cameras. Needless to say, nothing came of the hunt.

The same team returned to Sumatra in May of 2004. This expedition was to concentrate on the "lost valley." Debbie had told us of this mysterious place on our previous visit. Situated beyond Gunung Tujuh, it had never been penetrated by explorers.

We were to meet up with Mr. Subandi, the owner of a small rural hotel. We had stayed with him briefly in 2003. He was a keen birdwatcher and naturalist. He had found some orang-pendek witnesses for us to talk to. The witnesses lived less than an hour's drive away in a village called Te Uik Air Putih. By a remarkable stroke of luck, a specimen of the titan arum (*Amorphophallus titanum*), the world's largest flower, was blooming in the same area. The titan arum blossoms only once in ten years, so this was an unmissable opportunity.

The village backed onto an area called "the garden," cultivated land that is used for growing crops. The garden merges seamlessly with the jungle, and in some areas is very overgrown. Due to its more open nature, one usually encounters more wildlife in the garden than the jungle proper. The titan arum is truly the Godzilla of flowers, and looks like some strange surrealist sculpture or something made by the BBC special effects department. It stands seven feet tall. The elephant's foot of a stem widens into a barrel-sized green bowl. This in turn flares out into the petal, which looks like nothing so much as a Spanish Flamenco dancer's red dress. Finally, a phallic spadix of bright yellow rises from within the petal's folds.

We found the house of the witness and interviewed him via Mr. Subandi. His name was Seman. He was a middle-aged man with a young child. Seman had seen the creature in an area of land adjacent to a river at mid-day in February 2004. Back then the area was overgrown. The creature was only visible from the waist upward. He estimated it to be two feet, six inches tall, but when we looked at the area ourselves, it seemed that the animal must have been over three feet tall. The height he indicated with his hand looked like three feet as well. The animal had short black hair, a broad chest with pink skin visible on it, and a pointed head, possibly indicating a sagittal crest. The ears were long. The creature vanished, and Seman said that he had the feeling it had fled to the river and swam across it, though he did not see this. The river was a torrent when we were there, but in February it was much lower. The animal had been in view for three minutes.

On visiting the area, we worked out that the creature had been seventy-two feet away from the witness. Seman produced a sketch showing a powerfully built, ape-like creature with broad shoulders, long arms, and a conical head. At no time did it raise up its arms, as gibbons are wont to do on the rare occasions when they move about on the ground.

We returned to the same general area the next day to interview another witness. Ata was in his twenties and had seen his creature about three weeks after Seman. He heard a strange cry coming from the same area of the garden where Seman had his encounter. The noises began at ten in the morning. They were a loud OOOOHA! OOOOHA! sound. Upon investigation, Ata found himself only sixteen feet away from a strange beast. It was three feet tall and had short black hair. Its prominent chest made him think it was female. Its lower half was hidden by vegetation.

He noticed that it had large owl-like eyes, a flat nose, and a large mouth. It seemed aggressive, and Ata said he felt the hairs on the back of his hands rise up in fear. Ata produced a drawing of a muscular, upright creature with large round eyes. It lacked the pointed head of Seman's description.

Next day Sahar, his brother, and another guide arrived, and we left for a remote village called Kutang Gajha. We spent two days lost. Sahar was not au-fait with this area. By pure chance we stopped by a farmhouse. The people there said that one of their relatives, a man called Pak Suri, knew the way to the lost valley. Pak Suri was away that day and would not be back until the morning. The family kindly put us up for the night.

It transpired that Pak Suri would not be returning the next day as at first thought, but another man, Pak En, who knew the way, was contacted. Pak En was a sprightly old man who had ventured into the valley years ago on a fishing trip. He agreed to be our guide for the next few days.

Next day we set off, beset by leeches, flies, and biting ants. We daubed our boots in damp tobacco, as it drove the leeches away.

Finally, we came to the valley. There was a damn good reason why it was lost. Sheer cliffs fell one thousand feet into rapids. The sides of the valley were swathed in savagely thorned rattan. We had no rope. If we wanted to see the bottom of the valley, we would have to risk scrambling down by hand.

Pak En found a part of the valley wall that was slightly less than perpendicular, and we gingerly began our descent. What looked like solid ground would often be no more than a loose topsoil of leaves and would cascade from underfoot. Sturdy-looking branches would be rotten to the core and snap whilst being used for support. Half-sliding, half-walking, we made our way toward the bottom.

Walking out into the sunshine of the valley, it was astounding to think that I was the first Westerner ever to set foot in the place. It was more of a river-carved gorge than a valley. The fast-flowing river dominated the area. Though not deep or very wide, it was fast, and its bed was a mass of slippery rocks. The only place large enough to build our camp was in a small area of jungle close to where we had descended. The river looked as if it could flood violently and quickly.

At camp that night, Pak En told us that he had seen an orang-pendek in the jungle just above the valley three years ago. He was walking along a jungle trail when he saw it approaching. It was three feet tall, upright, and powerfully built. It had black hair with red tips and a broad mouth. Its prominent breasts made Pak En think it was a female. He noticed that it grasped the vegetation as it moved. It let out an OOOOHA! OOOOHA! sound. He watched it move down the trail for two minutes before it saw him. On seeing Pak En, it quickly turned about and walked back the way it had come.

That night the camp was eerily lit up by thousands of green fireflies.

After breakfast Sahar, Jon, Chris, Pak En, John, and myself set out to explore the valley. The nature of the valley compelled us to keep crossing the rapids on foot. The banks would peter out into sheer cliffs on one side forcing us to cross to the other. Some areas of the cliff faces were stripped clean by landslides. Hundreds of tons of earth, rocks, and trees had fallen into the valley, blocking whole areas and making the journey more arduous.

We had to scramble across slick boulders and walk across fallen trees. Such was the environment of the valley that it took hours to walk a distance one could have done in thirty minutes in England.

The progress was so slow that we realized that we would not make it to the end of the valley and back to camp before nightfall. We had to turn back about three-quarters of the way along the valley. Darkness falls with alarming rapidity in the tropics. The river was treacherous enough by day; in the dark it would be deadly. A broken leg in such a remote area could mean death. Sadly, we turned and headed back to camp.

We decided that, from where we were camped, it would be impossible to reach the end of the valley in a day. The small area was the only part of the valley suitable for camping. We had no choice but to climb up the cliffs to the top again. The valley did not look like suitable orang-pendek habitat. It was too narrow, and there was nothing in it worth expending all the energy of climbing down for. I think orang-pendek would have more common sense than to climb down into the gorge.

The climb back up was easier than that going down. We could crouch on all fours, making ourselves more stable. Once we had reached the top, we found a new place to make a fresh camp, and Pak En took us off to where he had seen the orang-pendek. It was a long climb up through harsh jungle. Along the way we saw scrape marks left in the earth by a tiger. It was odd to think that we were sharing the forest with such large predators. It is a feeling one seldom gets in Britain. Some people we spoke to had lived their whole lives in the jungle and had never seen a tiger. Sahar had only ever seen one. Mr. Subandi had seen a total of three.

When we reached the area of the sighting, Pak En mimed the strange way it was walking, gripping at the plants as it went. He told us that its outsized muscles reminded him of Mike Tyson.

After returning from the jungle, we headed for Banko, a town in Jambi province from whence we would enter the lowland jungles. Here we were to meet with the Suku Anak Dalam, the Aborigines of Sumatra, and interview them about the orang-pendek.

Sahar knew a man in Banko who spoke the language of the Suku Anak Dalam and could translate for us. We set out the next day, together with our translator, for a bumpy ride along an ill-maintained road into the jungle. The Suku Anak Dalam were once a totally nomadic tribe. Their only weapons were spears. They did not even use blowpipes or bows. They are far taller than the average Sumatran (who are of Malayan descent), with curlier hair. These days they are semi-nomadic, spending months in the jungle, then returning to live for a while in houses.

We found a chief, a man named Nylam, in a roadside house with his family and several members of his tribe. He had been suffering from malaria and was glad when I was able to give him some medicine. He seemed happy to take us into his house and speak with us.

Nylam had seen an orang-pendek in the area only three months ago. He had been up a tree at the time. The animal was four feet, nine inches tall and covered with red-tinted, black hair. It had a broad mouth. It walked upright and held its arms like a man. It made a WEEEEHP! WEEEEHP! noise and looked about itself as if it could smell its observer. Nylam watched it for half an hour.

Soon we had to return to England, and it was to be five years before I returned to Sumatra.

Chris Clark joined me again, and there were two new faces. Adam Davies was a fellow explorer and cryptozoologist. The year before, I had been to Russia with him in search of the almasty. He had visited Sumatra on several occasions before. Dave Archer was a CFZ member who had also been on the Russian trip and was keen to come with us to search for the short man.

Adam had contacted an Indonesian named Dally, who was to act as a "fixer" for us in Sumatra. Dally and Sahar met us at Padang airport. We were joined by John, Sahar's brother, who we knew from previous expeditions, and another guide called Doni.

For the first part of the trip, we planned to stay with the Suku Anak Dalam people that Chris and I had met in 2004. For the second half of the expedition, we were to return to Gunung Tujuh, the lake of seven peaks, the jungle-swathed crater of an extinct volcano in Kerinci Seblat National Park. There had been a number of orang-pendek sightings here in recent months, including one by an ornithologist.

After a long journey, we finally reached the rangers hut on the outskirts of town where we were to apply for permits to stay in the lowland jungles. There are only around two thousand Suku Anak Dalam left, so the Indonesian government are protective of them. Unfortunately, the head ranger was away in Java for two weeks, and we could not get permits to stay. However, we could visit the Suku Anak Dalam and speak with them.

We walked a couple of miles to a meeting point in the jungle. A small number of Suku Anak Dalam were waiting for us. This group seemed shyer than those we had met in 2004. The women and children ran away. There was only one man with the group. The other men were away hunting in the jungle. The man, who seemed afflicted with scabies around his feet, would not give his name but just told his story through a translator.

Three years previously, he had seen an orang-pendek close to the wonderfully named village of Anoolie Pie, some twenty-three kilometres away. It was around four feet tall and covered with black hair. The creature's face reminded the man of a macaque, with a flat nose and broad mouth. It stood and walked on two legs, never once dropping down on all fours. It was not a monkey, gibbon, or sun bear. The creature seemed afraid of him and walked quickly away whilst looking from side to side.

That evening we had a visit from an unassuming man called Tarib. He was the supreme chief of the Suku Anak Dalam. Most of his people were away hunting but he had made a special effort to visit us. He had an amazing story to tell.

Five years ago, he had seen an orang-pendek as he was walking in the forest. It was four feet tall, with black hair that shaded into blonde and grey in places. Its face looked like a monkey's, but it walked upright like a man. He took the creature by surprise, and it became aggressive. It raised its arms above its head and charged at him. He fled and hid behind a tangle of rattan vines. He watched as it looked for him, turning its head from side to side. Finally, it moved away.

This is one of the very few reports in which an orang-pendek acted aggressively. In all other cases, the creature has moved away quickly from the presence of humans. There is a story that, during the construction of the Trans-Sumatran Highway, the machines were attacked by groups of orang-pendek wielding sticks.

The next day we were up early for the long, hard climb up to the Lake of Seven Peaks. We crossed the lake again on the crumbling canoes that looked as if they would sink at any moment, then set up camp.

After breakfast, we split into two teams in order to cover more ground. Adam, Dave, Sahar, and Doni would take one track to a place where Adam had found and cast an orang-pendek track in 2001. Chris, John, Dally, and I would take another track, closer to the lake.

Dave had brought four camera traps and Chris had a number of sticky boards. These are actually methods of pest control. They are cardboard strips coated with a powerful adhesive that are laid out to trap rats and mice. We intended to place them on jungle paths, baited with fruit, in the hope an orang-pendek would leave some of its hairs stuck in the solution.

Our trail led for several miles alongside the lake. We came across some orang-pendek tracks. I had seen these before, and instantly recognized the narrow, human-like heel and the wider front part of the foot. They were impressed in loam on the forest floor and not good enough to cast. We set up two camera traps in the area and two sticky boards that we baited with fruit.

Upon returning to camp, we heard amazing news. Whilst walking through the jungle Adam, who is a mean tracker by his own right, had heard a large animal moving through the forest. In the distance, siamang gibbons were kicking up a fuss. Sahar and Dave crept forward and were greeted by an amazing sight.

Squatting in a tree around a hundred feet from them was an orang-pendek! They could not see the face clearly, as it was pressed against the tree trunk. Dave felt that it was peering at them from the side of its face. He saw the creature's eye rolling round in alarm and could see large teeth in the bottom jaw. The creature had broad shoulders and long powerful arms. The hands and feet were not in view. The orang-pendek had dark brown fur, almost black. Its texture reminded Dave of that of a mountain gorilla. This makes sense, as the jungles here are of a very similar type to those inhabited by mountain gorillas in Africa. The shape of the head recalled that of a gorilla as well, but the high forehead was like that of an orang-utan. The head lacked the long mane of hair described by some witnesses. He could see a line of darker hair running down the creature's spine. It was the size of a large male chimpanzee. He was sure it was not a sun bear or a siamang gibbon.

Next to the tree was some rattan vine the animal had been chewing. Adam carefully placed this in a specimen tube full of ethanol in the hope that some of the cells from the creature's mouth would have adhered to the plant, much like a DNA swab. A number of hairs were also found close by.

Next day, we checked the camera traps and sticky boards. The former had captured nothing, the latter only insects. We reset the camera traps and set out fresh sticky boards. We found some more tracks, but our plaster of Paris had degraded in the humid conditions and was useless. We continued our camera setting and hiking for several more days, with no results save for a picture of a small bird. In the afternoon, we decided to cross the lake and search on the far side. Adam had only been there once, and the rest of us had never seen the area.

The guides strapped the three canoes together with rattan to make a crude catamaran. The waters of the lake were calm, so we made the forty-minute crossing without incident. We waded ashore to steep-sided jungle slopes that seemed far less disturbed than they had on our side of the lake. The lack of areas to make camp meant that even the fishermen rarely visited this side of the water.

It was even damper on this side of the lake. Everything seemed spongy and had a rotten feel to it. We came upon a set of orang-pendek tracks that were clearer than any we had seen before. The toes were all individually visible. We photographed them extensively and cursed our lack of plaster to cast them.

On the crossing back, the waters cut up rough and the canoes almost sank.

After a short rest, we went and collected all of the camera traps and sticky boards. We returned to the tree where Sahar and Dave had seen the orang-pendek. We photographed Dally sitting in the same position. We also, under Dave's guidance, measured how large the upper part of the creature would have been from buttocks to the top of the head. It was three feet and four inches, making the animal, even if it had comparatively short legs, quite large.

After a long, long journey via Padang and then Singapore, we arrived home. I sent half of the samples off to Lars Thomas at Copenhagen University. Adam sent off his half to Dr. Todd Disotell of New York University. Lars studied the structure of the hair and found that it was similar to, but distinct from orang-utan. He commented that he was forced to conclude that there was a large, unknown primate in Sumatra. His colleague Dr. Tom Gilbert found some DNA that seemed to be human. We think the sample may have been contaminated during collection. Dr. Todd Disotell could not extract any DNA from his sample.

Shortly after my return to England, Dally emailed me twice to tell me of further orang-pendek sightings in Kerinci. On October 8, some bird watchers from Siulak Mukai Village saw an orang-pendek near Gunung Tapanggang. They watched it for ten minutes from a distance of only thirty feet. It had black skin and long arms. It walked like a man.

On October 18, a man called Pak Udin saw an orang-pendek in Tandai Forest. The creature was looking for food in a dead tree, possibly insect larvae.

It had black and silver hair, long arms, and short legs. He watched it for three minutes before it ran away.

In 2011, I was back in Sumatra as part of an international team. Adam, Chris, Dave, and I were joined by Andrew Sanderson, a veteran of a number of expeditions, Jon McGowen, naturalist and taxidermist, Tim de Frel, a Dutchman who worked for CITES (the Convention on the International Trade in Endangered Species) and Rebecca Lang and Mike Williams, who run the Australian office of the CFZ.

The group was to split into two teams, with Chris, Dave, Andy, and I taking the highland jungles and Rebecca, Mike, and Jon exploring the lowland jungles and the "garden" area. Tim would move between the two groups.

Before the teams split up, Sahar introduced us to a witness in his village. Pak Entis had seen an orang-pendek in April of that year in the garden area. He described it as three feet tall, with massive shoulders and tan-coloured hair. It had an ape-like face and walked erect whilst swinging its arms. Upon noticing Pak Entis, it became nervous and began to shake. It raised its arms above its head and made a "HOO-HOO" sound before moving away. It was in view for around sixty seconds.

Our team made the gruelling climb up the slopes of the mountain and the uncomfortable crossing of the lake to our camp area. Next morning Sahar took us into the jungle. On the first day, Sahar found and destroyed four snares set by poachers. It was the first time I had seen snares in the area, and demonstrated how human pressure is mounting in the park.

We set up our camera traps and found chewed rattan and fruit as well as a rotten log that had been pulled apart. There were no claw marks such as a sun bear would leave.

The following day, we moved to the far side of the lake where the ground is damper. Beside a rotting log that had been overturned by some powerful animal, Sahar found a handprint. The animal had braced itself with one hand whilst overturning the log with the other. It had torn the log apart looking for grubs. Andy quickly took a cast using plaster of Paris. Nearby, hair samples were found and taken.

The hand was 6 inches long by 4.5 inches wide. It had a rounded palm and thick, sausage-like fingers. The thumb was short and almost triangular. The print was nothing like the handprint of a Sumatran orang-utan, with its long thin fingers and palm and almost vestigial thumb. All in all, it looked more like the handprint of a small gorilla in configuration. The shape suggested that its owner would have been able to manipulate objects and use tools.

The rest of our days yielded little more, and the camera traps turned up nothing. We did, however, find the skeletons of two tapirs killed by tigers. Soon we descended once more.

Late in 2012, I was contacted by a French filmmaker called Christophe Kilian. Chris had worked on a number of projects for Scienti Films, a French company that specializes in science documentaries. Chris wanted to make a film about "wildmen" around the world, but focusing mainly on orang-pendek.

Chris had read my book *Orang-pendek: Sumatra's Forgotten Ape* and wanted to return to the jungles of Kerinci Sablat National Park with me to look for evidence of the creature.

The project was pencilled in for January, but was delayed until late June/early July, which, as it turned out, was a piece of serendipity. Joining us would be the beautiful and talented artist and taxidermist Adele Morse. I had met Adele in 2011, when she contacted me wanting information on orang-pendek. Adele had found my writings about the creature online and had decided to do an installation, or art project, on the creature for the Royal Academy. I gave her as much information as I could and sent her a copy of my book. She came up to Exeter for the weekend to interview me and subsequently made a number of life-size replicas of orang-pendek.

Sadly, Sahar had suddenly passed away just a few weeks after our 2011 expedition. Dally, acting as a fixer, had contacted Sahar's brother Jon and his son Raffles to act as guides. At Polempek, Sahar's village, the fixers had gathered a group of witnesses for us to interview. It was, to my knowledge, the largest group of orang-pendek eyewitnesses ever assembled. All of the men were locals and had seen the creature or its tracks in the area within the last year. With the fixers translating, Adele and I interviewed the men whilst Chris filmed us.

- Herman Dani had seen the creature at Uhan Danda a year ago. He only got a good look at the head. The face had a flat nose and thin eyes. Its fur was grey. The creature stared at him, and he ran away.
- Salim had seen strange tracks in the jungle five months before. He said they were man-like and the size of a human hand.
- Amri had his encounter at Padutingi, about four hours from Polempek, seven months previously. The creature he saw was one metre tall with grey fur. He ran away in fear.
- Juha Rapti had come across prints eight months before at Sungi Minya, around four-and-a-half hours away. They were human-sized, but had a highly separated big toe.
- Rahman saw the orang-pendek five months past at Gunung Sanka, about one hour away. The creature was large with black hair that faded to grey. It was moving quickly, and he didn't get a good look at it. He fled.
- Saba Rudin saw the creature as it crossed a jungle trail. The area was between Sungi Mina and Sungi Kuni, about five hours away. The event had happened ten months before. The orang-pendek had a broad, barrel-like chest and black and grey hair. It walked on two legs like a man.
- Aprisal was in the Sungi Kuni area nine months before, hunting wild pig. When he paused, he saw a creature with black and grey fur, and a large mouth. Afraid, he ran away. The area was about four hours distant.
- Mah Darpin saw the orang-pendek after rainfall at Gunnung Kacho, nine hours distant. He saw the creature from the back, noting that it had long fur of a grey-black colour, was around three feet tall, and walked on two legs. He became afraid and walked back the way he had come.
- Saimi Alwi saw the creature one year before at Sungi Minya, which is about four hours from Polempek. Whilst tracking, he heard a noise and saw a creature squatting to eat kitan fruit. The animal was barrel-chested and muscular. It had black and grey fur and stood three feet tall.

Their stories were remarkably consistent, and I was struck by the lack of exaggeration.

The next day we began the arduous climb up the rim of the extinct volcano that forms Gunung Tujuh. After resting at the top, we descended to Gunung Tujuh itself. I was horrified to see that the boulders at the edge of the lake had been defaced by graffiti. It seems that true wilderness is getting harder and harder to find.

We made camp. Adele, Chris, and I had our own tents whilst the guides and fixers erected a pondok, a large shelter made from branches, palm leaves, and plastic sheeting. As the sun was setting and a fire was being started, a primate call unlike any other I have heard rang out from the forest. I am familiar with all the calls of the wildlife in this area. I have also kept all the local primates in captivity. This was the call of a primate, but one I had never heard before. The vocalization went, "ho...ho...ho...ho."

The guides and porters froze, and John Dimus said "orang-pendek." Chris frantically tried to record the sound, but it had ended by the time he had his equipment ready, and the call was not repeated. It had sounded relatively close to camp—an exciting start to the expedition.

In the morning, we trekked into the jungle. There was much evidence of tiger activity. Within a mile of the camp, we found tiger claw marks on a tree trunk. We came across two tiger kills. Both were Malayan tapir and the bones had been picked clean. Adele took a coccyx and vertebra as well as some teeth for her art display.

It was late June, and a fruit called the kitan by locals was ripe and falling. About the size of a pineapple with a reddish-brown, hard outer skin and a yellowish pulp inside, it is said to be favoured by the orang-pendek. I have not yet been able to find the scientific or Western name for the fruit. The guides said that only the orang-pendek favours it. We found several that looked as if they had been chewed by something with human-like teeth. Close by, John found a hair, but touched it with his fingers! Still, we preserved it in ethanol.

We set up a camera trap and baited it with fresh mango.

The following day, we decided to explore the far side of the lake and set out in the rickety canoes. We climbed up a ridge and listened to siamang gibbons calling. Whilst the rest of us were listening to the gibbons, Adele came across a track. It was on a slight slope, but it clearly showed the human-like heel and broader front foot typical of an orang-pendek.

I had brought some dental cement with me. This is a kind of fine plaster of Paris ideal for casting tracks. I had tested the substance out in my garden back in Exeter to see how much was needed to make a cast and to practice until I got the consistency correct. As the substance was heavy, I carried only as much as I needed to cast a couple of tracks.

Adele offered to make the cast, as she was a sculptress. However, I had not reckoned with the greater moisture in the rain forest and the more porous consistency of the ground. The casting of this one track took up all the dental cement that I had brought. Adele tried to strengthen the cast with strips of cloth. Once it was dry, we gingerly eased it from the damp earth. It was not the best track I have ever seen, but the heel and toes were visible. The detail was somewhat blurred by the action of the bunching up of the earth around it on the slope. The cast cracked in two, but Adele offered to take it back to London and repair it.

A few minutes later, we came across two handprints that looked very like the one found by Andrew Sanderson on our 2011 expedition. I was gutted that I had run out of dental cement. On future expeditions I will know to take far more than I think I need.

The following day, we followed another trail in the jungle in a different direction. Half a mile from the camp we came upon another orang-pendek track close to a rotting log. We also found and collected hair samples in the area. Whilst taking the samples we heard the call again. "Ho…ho…ho…ho." Once more Chris was foiled in his attempt to record the vocalization.

We moved further into the jungle and found another rotting log. Beside it was a set of the most perfect orang-pendek prints I have ever seen. No more than a day old, they were perfectly preserved in the damp soil. They clearly showed the long, man-like heel, four toes at the front, and the offset big toe at the side. They would have made amazingly detailed casts. Once more I cursed at not having brought more dental cement. I had to make do with filming and photographing them.

I believe that there are two keys to finding orang-pendek in Kerinci Seblat. Firstly, to visit when the kitan fruit is ripe in late June/early July. Secondly, rotting logs. There have been a number of reports of orang-pendeks seen ripping rotting logs apart to find grubs and insect larvae. We found the prints of at least three individuals around rotting logs.

In the same area we found many hair samples, far more than on any other expedition. I filled all my specimen jars and preserved the samples in ethanol. Again, we heard the now familiar call "Ho...ho...ho...ho" but further away in the jungle depths. We found chewed kitan fruit in the area as well.

On the way back to camp we were confronted by a mass of vegetation, a labyrinth of branches made up of both dead trees and fallen trees that were still alive. It was a hard climb across, round and through the jungle maze. Adele did particularly well despite the fact she has poor depth perception due to her lacking certain eye muscles.

The following day we set out to collect the camera trap. On the way we found more tracks, but these were not so clear as the ones near the rotting log. At the camera trap area, we saw that the mangoes had remained untouched. On our way back we faced a second vegetation labyrinth like the one from the day before.

Back in camp we looked at the shots from the camera trap. It showed nothing except us putting it up and taking it down. I was not surprised. When testing the traps back in England, we found that they had to be left up for weeks on end before they captured anything at all.

Eventually we made the long crossing back over the lake and climbed down the side of the crater. The crossing and trek took most of the day. We saw monkeys and pig hunters as we reached the lower forests. At Polempek the cars were waiting to take us back to Sungi Penuh.

We met up with Dr. Achmad Yanuar of the National University of Java. A primatologist, Dr. Yanuar had worked with Debbie Martyr and Jeremy Holden during the Flora and Fauna International orang-pendek hunt in the 1990s.

Dr. Yanuar unfurled a map of Sumatra and showed us all of the places he had hunted for the orang-pendek. He had investigated sightings and findings of tracks over most of Sumatra except the north. In all of his years searching, he never saw the creature himself or found any tracks, but he interviewed many witnesses. It seemed that even twenty years ago, reports of orang-pendek were much more widespread. Today they seem to be mostly confined to West Sumatra and Jambi. Though its range has shrunken vastly and swiftly, the creature seems to be hanging on in these areas. We saw the tracks of at least three individual animals in one small corner of Kerinci Seblat National Park, an area that covers 13,791 km.

Whilst online at the hotel, Adele found a news item of extreme interest. Dr. Tom Gilbert, a geneticist from the University of Copenhagen, with whom we had worked with before, had begun an exciting new project. Dr. Gilbert was looking into extracting DNA from the blood in the guts of leeches to find out what animals they had fed on. Dr. Gilbert and his team got twenty-five leeches from the Annamite Mountains of Vietnam and successfully extracted mammal DNA from twenty-one of them. Mammals identified included the striped rabbit (discovered in 1995), the serow (*Capricornis sumatraensis*), the Chinese ferret-badger (*Melogale moschata*), and the Turong Son muntjac (*Muntiacus truongsonensis*), which was only discovered in 1998. The DNA can remain intact in the leech's gut for several months. He went on to say that the technique could be used in searching for creatures such as the thylacine and the orang-pendek. We intend to work with Dr. Gilbert and try his leech method on future expeditions.

The following day we travelled to Banko. We wanted to find some lowland forest to interview Dr. Yanuar in. However, the lowland forests were so degraded it proved hard. Most had been chopped down to make way for palm plantations to provide palm oil. We checked into a hotel and then went out to talk to one of the Suku Anak Dalam.

We drove from Banko to a small roadside village to meet a Suku Anak Dalam man called Pak Tumcuggung. Pak Tumcuggung had recently converted to Islam and lived in a house, eschewing the old ways of his people.

He told us, via a translator, about his encounter. About forty years before, he had been walking through a graveyard about a mile from the village of Batanlumbhi. At the time, the area was heavily forested. Today the jungle has been cut down to make way for palm oil plantations. It was the rainy season and about five in the evening. Pak Tumcuggung came upon some odd-looking tracks. Then he saw a man-like, grey-coloured figure rise up from behind one of the grave markers. At the time he thought it was a ghost and referred to it as orang-hutan or ghost man.

The creature stood around three feet tall. It had long grey hair, broad shoulders, and a pot belly. The face looked very human, with broad cheekbones. The creature looked more like an orang-utan than a siamang gibbon. The two stood and stared at each other until Pak Tumcuggung turned and ran. He looked back and saw it still standing there watching him.

At the time, he thought what he had seen was a ghost, because he saw it in a graveyard and it seemed to combine animal and human features. Now he realizes that he saw some kind of animal. He feels it is related to the orang-utan, but lives on the ground rather than in trees. He feels it still exists, but not in the same area, as that has been deforested.

His brother saw an orang-pendek the same year. It was very like the one Pak Tumcuggung described except it had black hair rather than grey.

In 2010 local people heard strange calls from an area of forest. It was not a siamang. He thought it may have been an orang-pendek.

I intend to return to Sumatra, as I think the orang-pendek is on the cusp of discovery.

The Tasmanian Wolf

If there is one creature I am even more convinced of than the orang-pendek, it must be the Tasmanian wolf. We have already looked at the case for its survival. The Centre for Fortean Zoology chose this creature as its logo and emblem because it is an icon for both cryptozoology and conservation.

I had always wanted to go in search of the thylacine. Back in the '90s as a student, I spent some time planning an expedition, only to be lied to and have my research stolen by a UK film company—something that I am still bitter and angry about.

When the Centre for Fortean Zoology's Australian representatives, Rebecca (Ruby) Lang and Mike Williams, came up with the idea of an official CFZ expedition to Tasmania, I was thrilled to be on board. Mike and Rebecca were to be joined by another old friend of the CFZ, veteran Australian cryptozoologist Tony Healy. Tony had spent a lifetime on the track of unknown animals all across the world. Also joining us would be Tania Pool, a CFZ member and researcher who had joined us at the Weird Weekend, the CFZ's annual convention, and the Fortean Times Unconvention on a number of occasions. Rebecca's friend Hanna would round off the Australian team.

The British contingent left Heathrow on October 31. It felt truly special to be searching for the creature so iconic that the CFZ adopted it as its logo and totem animal.

I am used to long flights, but the trip to Tasmania was something else. We stopped in the Middle East, Borneo, and Melbourne before reaching Launceston in Hobart over twenty-four hours later. We were met by our Australian friends, and in no time we were driving to Launceston. Tasmania is alive with wildlife, and on our first relatively short journey we saw common wombats (*Vombatus ursinus*), Bennett's wallabies (*Macropus rufogriseus*), and a short-beaked echidna (*Tachygossus aculeatus*).

Launceston, Tasmania's second city, is the size of a medium town in the UK. It has an Old World feel about it with colonial-era buildings and houses. The Tasmanian wolf is everywhere, on car registrations, in shop signs and council logos. The creature is very much alive in the island's iconography. Two of them even appear as supporters on the Tasmanian coat of arms.

Moving on inland we reached the small town of Mole Creek in central Tasmania. We had been booked into the Mole Creek Hotel, home of the famous Tasmanian Tiger Bar. The hotel itself has a sort of old-fashioned charm with a 1950s feel to it. I felt very comfortable there and it had a lovely atmosphere. The Tasmanian Tiger Bar is like a small museum filled with thylacine memorabilia. There are paintings, sculptures, and a well full of framed newspaper reports of sightings. They even serve Tasmanian Tiger Ale, a very tasty pale ale that I imbibed several pints of.

The landlord, a charming man called Doug Westbrook, was good enough to give us an interview. He showed us alleged droppings (desiccated and in a jar) and a number of prints. The prints did indeed match those of a thylacine rather than a dog, wombat, fox, or any other animal. Doug himself had never seen the Tasmanian wolf, but his wife Ramona had. Ramona did not like to speak about her experience, so Doug gave us the details. In 1997, she was driving along a country road about sixteen kilometres from Mole Creek when a thylacine loped across the road in front of her. She noted the striped rump and stiff tail. She described the creature's gait as "awkward" and said that its hindquarters seemed to move stiffly. "Like a dog with a broken back" had been her description. As the animal reached the far side of the road, it turned its head back to glance at her. Then it was gone into the bushes.

Doug said that in 2011, a French girl staying at the hotel had a very similar sighting. About two kilometres from Mole Creek, she also saw a thylacine crossing the road in front of the car she was driving. As with Ramona, she saw the striped rump, stiff tail, and strange gait. She too had used the phrase "like a dog with a broken back" to describe how the animal had moved.

We visited the Trowunna Wildlife Park close by. As well as birds, reptiles, kangaroos, wombats, and echidnas, the collection includes the island's other marsupial predators, spotted quolls (*Dasyurus maculatus*) and Tasmanian devils (*Sarcophilius harrisii*). Up close, the resemblance of the quoll's face to the thylacine's is striking. Both have dark eyes, rounded ears, and a dog-like snout. There the resemblance ends. The body and tail look more like a stout cat with brown fur and cream spots. The bulkier Tasmanian devil looks more like a hybrid of bull terrier and badger. The devils are currently beset by a form of transmittable cancer that affects the face of the animal. First seen in the mid-1990s, the disease causes huge facial tumours that lead to death. Devil facial tumour disease has caused a population crash of 50 percent. Trowunna maintain a large, healthy breeding population in captivity as a safeguard against extinction in the wild.

We travelled in two magnificent Toyota land cruisers, kindly loaned to us by the company, and Tony Healy's trusty old Volkswagen van. On the road, Tony unveiled his maps like wizard's spell books. They were dotted with notes and annotations in astonishing detail. Arrows and dots pointed out locations and dates of sightings, not only of Tasmanian wolves but of bunyips, sea serpents, yowies, and even ghosts as well.

We headed out to the Cradle Mountains and the Wilderness Gallery, where there was an impressive and shocking exhibition on the thylacine. It featured a thylacine skeleton, pelts, skulls, and a reconstruction of an old trapper's hut from the nineteenth century. On display was a book logging the captures and killings of thylacines from the bounty era. It covered the late nineteenth and early twentieth century. It felt odd to be actually touching the book, with its original notations of specimens, locations, and payments. The numbers being brought in dropped sharply in the twentieth century.

The main exhibit was the Tiger Buggy Rug, an object both appalling and fascinating—a carpet made from the hides of eight thylacines. There was also film on a loop showing the last captive thylacine wandering around its barren enclosure at Hobart Zoo. I've seen the film many times before, but it was intercut with other, older, rarer clips of captive animals.

Annoyingly, the whole exhibition focused on thylacine extinction. There was not one word about thylacine survival or any of the four-thousand-plus sightings since 1936.

Prior to embarking on the expedition, we had all agreed to keep the focus area a secret. Therefore, I will not reveal where we did our work, other than that it was in the northeast of the island.

On the way we saw much wildlife, including another echidna and the ubiquitous Tasmanian native hen (*Tribonyx mortierii*). The birds, which are actually flightless rails, are found just about anywhere there is water. Our campsite was a small affair off an old logging road. There was an organic toilet, a couple of brick barbeques, and a sheltered area with tables for eating.

We set up camp without delay. Tony was sleeping in his van, the rest of us in tents. CFZ stalwarts Jon Hare and Chris Clark had two small tents of their own. Mike and Rebecca shared a larger one and taxidermist Jon McGowan, Tania, and me shared Tania's huge tent (that she had picked up at a secondhand shop).

The camp had its own residents. One was a noisy brush-tailed possum (*Trichosurus vulpecula*) that disturbed the night with vocalizations one would not believe such a small and endearing animal could make. The second was a black currawong (*Strepera fuliginosa*), a yellow-eyed, crow-like bird always on the lookout for scraps.

The forest floor around the camp was studded with what looked like huge worm casts. We dug down into the earth to try and find what we believed to be the massive worms that created them. We had no luck.

That night we conducted the first of our night drives. We had cameras mounted on the hoods of the land cruisers and Tony's van. These were left running throughout the drives in case our target animal should run across the track in front of us.

The area we were searching in was remote. It was heavily forested and away from main roads. We followed old logging tracks, some unused for years. The forest was thick with wildlife, all of which would make fine prey for the Tasmanian wolf. On the first night alone we saw Bennett's wallabies, red-bellied pademelons (*Thylogale billardierii*), a form of small wallaby, wombats, and a Tasmanian spotted owl (*Ninox novaeseelandiae*).

Despite being the beginning of summer in Tasmania, it was still bitingly cold at night.

Next day we took a trek to a local lake, a large, water-filled sinkhole. Around this particular lake, people have claimed to have heard the distinctive call of the thylacine at night. It is said to be a high-pitched yap in three parts: "yip-yip-yip." It said to be quite distinct from all other native animals, and quite unlike a fox's.

We set up some camera traps, sensitive to both heat and motion. These we baited with leftover chicken, cat food, and bacon jerky. We set up other cameras along a long closed and barricaded road, reasoning that this would be doubly undisturbed. We searched for roadkill to use as further bait but found none.

Along another one of these logging spurs, we found some large droppings. They were transparently those of a carnivore, containing as they did bone shards and hair. They seemed too large to be from a devil or quoll and too remote to be from a dog. We carefully preserved them in a solution of 70 percent methylated spirits and 30 percent water. Thylacines were reported in this area in 1994 and 1996.

The area had several small rivers running through it, and we came upon a bridge that was so rotten it would not hold our vehicles. We carried on, on foot.

Later we visited an open area of button grass. As darkness fell, we took watch. It was bitingly cold and perfectly still. Nothing moved, and we saw and heard nothing.

We came upon a dead chicken and hung it beside one of the old logging spurs with another camera trap facing it. Another night drive turned up more wombats, brush-tailed possums, pademelons, and wallabies.

The following day Tony took his van into town for a service. In the garage he met a woman whose father had seen a thylacine in the general area in the 1970s.

Rebecca, Hanna, and Jon Hare drove to town to buy some extra quilts and blankets.

Jon McGowan, Chris, Mike, and I searched for snakes with little success. After nightfall we went out on foot, spotlighting for animals, and again saw much fauna.

We drove down to Allendale Gardens in the morning. These open gardens are a collection of beautiful landscaped areas and natural woodland featuring gigantic, ancient trees that were saplings when Europeans first discovered the island. The landlord, a tall, bearded man named Max Cross, had seen a thylacine in 1996 and he was good enough to grant us an interview.

He had been driving between Hobart and Launceston when a large thylacine had rushed across the road. Once again, the stiff-looking, striped hindquarters were emphasized. It was the size and general shape of a large dog and he pointed out his own dog Myska, a large crossbreed, as a good comparison. He also mentioned a stiffly held, thick tail. Max noticed how, after crossing the road, the thylacine was moving up and down looking for a way through. Another car behind Max's slowed down and saw the animal. The occupants of the second car must have reported the sighting, as a story about it subsequently appeared in the *Launceston Examiner*.

Often, on our expeditions, we turn up information on cryptids other than the one we are actually looking for. This trip was no exception. We were talking to Max about Tasmanian wildlife in general and he mentioned that, when he first moved into the area, something had been killing his chickens. He shot the offending predator and it turned out to a spotted quoll, but one of mind-boggling size. This animal is usually around seven pounds in weight and around three feet long. Max said the animal he shot was the size of a cattle dog. This breed of herding dog is slightly larger than a border collie, weighing up to forty-nine pounds. He indicated the tail and body length by raising his hand from the ground to just above his shoulder. Max was a tall man, over six feet, so the length of the giant quoll would be over five feet. Max commented on the thickness of the animal's neck. In comparison, the thylacine averaged six feet in total length, with some larger individuals ranging from seven to nine and a half feet long. At the time, he had no idea of the value of such a specimen and threw it away. This occurred in 1964.

Later that day, we spoke to another witness. Damien Key was also impressively tall and impressively bearded. He worked in a family garage/shop, but he is also a government licenced shooter who is paid to keep the number of wallabies in check. He goes deep into the bush, shooting his prey and leaving the carcasses to feed the Tasmanian devils, and perhaps other things as well. He also culled feral cats that, as on the mainland, have proved a horrible menace to smaller native animals.

In 2008, in the area where we were based, Damien saw a large, dog-like animal run across the road in front of him. He noted the stiffly held tail but could not recall stripes. However, there are other records of stripeless thylacines. Most striped animals have variation in their markings. For example, there are tigers with very faint stripes.

Then, in 2010, he and a friend saw another thylacine a few miles from his first sighting. This time he did see the stripes, as well as the stiff tail and odd hindquarters, as it ran away into the bush. The following year he was approached by a logger who asked him if he had ever seen a thylacine. When Damien said he had, the logger confessed he had too, in broad daylight and in the same area. The man had been on foot and walking along the logging spur when he saw the thylacine.

Damien had also heard the distinctive call of the Tasmanian wolf on several occasions. It was distinct from a fox's and more spasmodic. One of the places he had heard it was a small, remote airstrip on which wallabies grazed at night.

Later that day, Jon McGowan came across a roadkill bandicoot (*Perameles gunnii*). He cooked and ate some of the creature. Back in England, Jon, who works at the Bournemouth Society of Natural Science, lives almost entirely on roadkill, feeding his guests badgers, foxes, and other strange delicacies. He offered me some smoked bandicoot flesh, but it smelled a little rancid for me. The rest of the carcass we used to bait a stream close to our campsite, hoping to attract the Tasmanian giant freshwater crayfish (*Astacopsis gouldi*). This is the world's largest freshwater invertebrate, weighing up to eleven pounds and being nearly three feet long. Unfortunately, despite baiting several parts of the stream, we never did attract one.

The next day, a wildlife guide and a couple of elderly tourists turned up. The man told us that the creatures that caused what we thought were huge worm casts were in fact burrowing crayfish. They have networks of burrows and shafts running up from the water table, and play a major role in soil turnover, drainage, and aeration. The guide said that he too had heard the distinctive yip-yip-yip of the thylacine on two occasions.

The following day was Jon McGowan's birthday and he received not only a large cake but the gift of a roadkill Bennett's wallaby. The creature had been found by some of the team when they had driven to a small town that morning. Jon gleefully skinned and butchered the creature. We used some of the meat to rebait a number of the traps. We also took the opportunity to look at the images we had captured so far. The camera from the lake showed a tawny frogmouth (*Podargus strigoides*), a large nightjar-like bird, Tasmanian devils, and a spotted quoll. The camera from the logging road showed devils and a feral cat. Jon Hare had spoken to a woman at the café in town who had seen a Tasmanian wolf crossing the road close to the area in which we were camped. Unfortunately, he did not get any further details. Tony Healy had spoken to another witness, who worked at the town garage. He had seen a thylacine in 1984 at a creek about eight miles from the town. He felt that bush fires may have forced it out of its usual haunts. He and a friend had been on a motorbike at about ten thirty at night. They came across a large, dog-like animal. They noted that it had a long, stiff tail and striped hindquarters. The hindquarters looked large and awkward. When the animal moved, it swung them from side to side in a manner that recalled a cow. The movements were not like a dog's. They slowed down to look at the creature, and it took a couple of steps toward them. Becoming scared, they turned around and drove away. A woman at the garage also told Tony her father had twice seen thylacines in the 1980s, one at a local creek and another on a certain length of road.

That night we barbecued the wallaby. Even after using the head, tail, feet, and innards as bait, there was a large amount of meat left. It proved to be palatable, if a little tasteless. We visited the remote airstrip where Damien said he had heard the thylacine's call. It was empty, and the small Portakabin was locked and did not look like it had been used in some time. The strip itself had been grazed by wallabies, wombats, and kangaroos. We also visited the roads where thylacines had been reported. They were exceedingly overgrown. It was obvious no one had been along them in some time. Tony and I saw a magnificent and deadly Tasmania tiger snake slithering across the road in front of us. We leapt from Tony's van but were too late to catch up with the reptile.

We returned to the airstrip at night to stake it out. We heard a wallaby give a thumping alarm in the manner of an overgrown rabbit. We also heard owls. The night sky was punctuated by shooting stars and the rays of the aurora australis, or southern lights. Any thylacines lurking in the shadows remained silent.

Next day, Tony travelled back to the town and we caught up with him later. We met up with him for lunch in a pub. A local man, Mick, told us he had seen a thylacine at Serpentine Creek, an area quite some way from where we were searching. He was walking on a path off the river, and he saw the animal crossing the path ahead of him in broad daylight. He had been struck by the stripes on the flank. This occurred in 1988. Tony interviewed another man whose brother, a truck driver, had seen a thylacine in 1993 whilst driving between Scottstown and Georgetown in the northeast of Tasmania.

We tried to visit a small museum in the town to see if it had any material about the Tasmanian wolf, but it was closed.

Later we took down the camera traps and reset them with fresh bait at the airstrip and on the roads and hills around it. Rebecca and Hanna spent the night at the area where zoologist Hans Nardding had an excellent view of a large male thylacine in 1982. After a freezing night in the land cruiser, they saw nothing.

Next day we drove up to Mawbanna, where the last known wild thylacine was shot by Wilf Batty in 1930. Rumour has it that six more were caught alive in the general area in the 1930s, and one in the 1960s. These days the place looked quite unsuitable, being mostly cleared farmland.

That night at camp, Jon McGowan returned in a state of excitement after wandering in the surrounding forest. Opening up his hand, he showed us his prize with all the enthusiasm of a schoolboy. It was only a deadly funnel web spider with venom quite capable of killing a person. Rebecca, being an arachnophobe, was appalled as Jon played with the huge arachnid as if it were a pet mouse!

We tried to drive to Lake Rowalla, where thylacine tracks had recently been found. However, the sat-navs malfunctioned and we got lost. We ended up in the town of Waratah and had lunch at the excellent Bischoff Hotel, a magnificent building dating to the 1900s. Inside was a preserved specimen of the Tasmanian giant crayfish. It was quite as large as a big marine lobster. The owner told us of a local family who saw a thylacine crossing the road in front of their car near Rapid River in the 1970s. Apparently, they disliked talking about it.

At the small local museum, a 1970s newspaper was on show. It had a two-page spread about the Tasmanian wolf and detailed several sightings.

On the way back, we stopped to explore more remote logging roads. On one we found more large droppings from a carnivore and took them as samples. Close to camp, a spotted quoll bounded across the road in front of us and we caught it on camera along with several Tasmanian devils.

Rain marred the following day. Exploring the woods around the camp, we discovered a cave. We rigged up some ropes and, one by one, lowered ourselves into it. Spindly cave spiders with a leg span like a human hand lurked in the cave. Some sat upon egg sacs as large as hen's eggs. *Hikmania troglodytes* is the biggest spider on the island and belongs to a primitive group that is ancestral to modern spiders. Its closest relatives live in Chile and China.

Australian TV channel ABC wanted to do an interview with us. As we were nearing the end of our expedition, we agreed. We met them in a small seaside town. Whilst waiting in a café to meet them, I picked up a magazine and found an article in it that was an almost word-for-word ripoff of one I had written years before for the CFZ journal *Animals & Men*. It was about the creature known as the Gurt Dog of Ennerdale that terrorized the British Lake District in 1816. Its description and habits recalled a Tasmanian wolf, and I theorized that the creature had escaped from one of the horse-drawn, travelling menageries that were popular at the time.

Finally, they arrived. There was a likeable cameraman in his fifties and a young, somewhat pushy female presenter. The presenter wanted to film us discussing the expedition over a drink in a pub. We went to a pleasant pub overlooking the ocean and she ordered us a round of drinks. We were duly filmed talking about the thylacine and our trip. Then they wanted to film us "setting up" our cameras. Of course, we were not going to let on as to the real location of our expedition. We would just recreate what we did in some nearby bushland. As we began to leave, we found out that the girl had left without paying for our drinks. We had to pay for them ourselves! We were quite annoyed by this, but things got worse. She was obsessed with Bigfoot, and kept saying "You have hunted for Bigfoot, haven't you?" I repeatedly told her that I had hunted for the yeti, the almasty, and the orang-pendek, but never for Bigfoot, to which she said, "I've heard you have hunted Bigfoot." Again, I reiterated that I had not and neither had the other team members. We were filmed setting up the camera traps and interviewed about the expedition. When the piece was finally transmitted, the presenter said that we had previously hunted for Bigfoot, despite what I had told her.

Later we found a recently dead roadkill spotted quoll. Jon McGowan later skinned the animal and cooked it. The meat was succulent and far better than that of the wallaby.

Whilst Jon Hare, Chris Clark, Mike, Jon McGowan, and I had been enduring this rubbish, Rebecca and Hanna had managed to get a look around the little museum in the town that had previously been closed. They found a photo of a stripeless thylacine. This throws an interesting new light on Damien's first sighting.

On our final full day in Tasmania, we had a remarkable stroke of luck. Tony and Mike were talking to some folk in a café in the small town we visited. We had eaten in the café on a number of occasions. As it turned out, one of these men was Granville Batty, the great-nephew of Wilf Batty, the man who shot the last known wild thylacine in 1930.

Granville was good enough to speak to us for some time. He still had the gun that had done the terrible deed all those years ago. He had sold the farm in Mawbanna where the drama had unfolded. Mr. Batty said that there were thylacines in the Mawbanna area up to fifteen years ago. Sightings had dropped off in that area since the 1980s due to the plantations being laced with poison. His father-in-law had heard a thylacine calling whilst he was fishing on the Arthur River. Granville thought the thylacine could well still be about. He said that if he had the money, he would search south of the Arthur River. He related hearing that thylacines were fond of eating birds and that the ones in London Zoo caught pigeons. He also said he had been told of them hunting seabirds on beaches. I myself have read of them catching sparrows in captivity.

We returned to England, and our antipodean colleagues to the mainland. The samples were sent off to Copenhagen University to be examined by our old friend Lars Thomas. Surprisingly, they turned out to be those of a Tasmanian devil. It must have been a specimen of huge size.

We had already made plans to return the next year. The small population and vast wilderness convinced me more than ever of the Tasmanian wolf's continued existence.

In February of 2016, I returned to Tasmania for my second attempt to find the Tasmanian wolf. The last expedition I had taken part in, back in 2013, had consisted of many people from Australia and the UK. On this occasion, Mike Williams of CFZ Australia had decided, wisely, to pare it down. This time it would be a skeleton crew of Mike and myself on the track of the legendary beast.

Flying to Tasmania is a long affair, taking a day or more and three changes. Getting to your destination is the only leg of an expedition that really worries me. Once I'm in the field, I'm fine. I finally got to Launceston and was met by Mike. Over a coffee, he explained some developments since I'd last been to Tasmania.

I'd missed the 2015 expedition due to a bout of gout and pneumonia. On that trip, the team had been in the northeast of the island, but met with less success then on the 2013 trip in the northwest. The area we had visited on that trip had been subject to savage bush fires. Multiple lightning strikes had caused fires to reduce much of the forests to ash. It's a natural process in Australia, but it rendered the area less than perfect for our purposes.

Mike had been in contact with a farmer in the northeast who said he had captured a thylacine on camera. The man, who wanted to remain nameless, had allowed Mike to look at the pictures after much persuasion, but would not allow copies to be made. Mike was convinced of the authenticity of the pictures, mainly due to one interesting feature. The creature in the two pictures had a shaggy winter coat. Most people do not realize that the Tasmanian wolf grew longer hair in the winter months. Most reconstructions of them show the animal with a short coat.

The first picture showed the creature side-on to the camera trap, and the second showed it turning away. The stiff tail and stripes were apparent. The farmer had placed the camera trap on a hill on his property for months on end. These were the only two pictures he had gotten over that period. He was cagey about showing them to anyone else or being interviewed.

Mike had heard of some recent sightings further south on the island and, with our former area being burned out, we decided to make this our HQ for the trip. As before, we decided not to reveal the exact location of the sightings in order to protect the animals.

We camped out at a grassy area with wooded hills on the first night. We found a dead Tasmanian devil on the road. It had apparently been hit by a car. There were no signs of the facial tumours that have been ravaging the population elsewhere.

We had brought camera traps with us. We affixed these to trees in remote areas and used roadkill as bait. In addition to this, we once again employed bonnet-mounted cameras that film constantly as we did our night drives. Most sightings of the Tasmanian wolf are made by motorists at night. Should anything run in front of our vehicle, it would be caught on film.

Once more, Toyota had generously lent us a four-wheel-drive car for our trip.

Next day, we crossed the Western Tiers, a beautiful escarpment studded with lakes. We made camp and set up cameras. By day we explored on foot, and by night we drove the remote country roads. Bennett's wallaby, pademelon, wombat, and eastern quoll (*Dasyurus viverrinus*) were all in abundance. The tiger quoll and the Tasmanian tiger were not nearly as apparent as they had been further north.

The following day, we drove to a small town to meet our first witness, Joe Booth. When Mike and I arrived, Joe was in his garage, trying out a homemade prosthetic hand that appeared to have been made from a sharpened curtain hook and an old aerosol can. Greeting us enthusiastically, Joe, who was an instantly likeable bloke, explained about his homemade hook. The year before, he had been out with his mate, who was a keen hunter. Joe had been standing outside his mate's car. On the back seat was a dangerous combination of loaded guns and excitable dogs. As the dogs bounded about, one of them knocked the guns, which had the safety catches off. One went off, blowing a hole through the side of the car. It also took a chunk out of Joe's side and blew off his right hand.

Joe was lucky to survive, and had to have several transfusions. However, he bore his mate no grudge and seemed to take his disability in his stride and did not let it affect him in the least. He found the prosthetic hand given to him by the hospital uncomfortable and got on better with the one he knocked up at home.

Joe had been a logger in the '50s, '60s and '70s and had seen some of the most massive trees in Tasmania fall to the power saw. He told us that in remote areas the crew would regularly come upon dog-like tracks. He asked the foreman who on the crew had a dog. The foreman replied that they were the tracks of a Tasmanian tiger. One of the other workers scoffed at the idea. A few days later, the same man walked around a large tree stump and found a thylacine sitting there. The animal gave a warning gape and the man backed swiftly away.

In the 1950s, Joe had his own sighting. One evening, he was putting his car away. It was twilight and he saw what he thought was his neighbour's dog walking down the road toward him. He called out to it, but it didn't react. As it drew closer and walked past him, he saw it had a thick, stiff tail and stripes along its hindquarters. He then realized that he had seen a thylacine. He had recalled hearing that a crop sprayer pilot had said he had seen one in the vicinity some days earlier. A few days after Joe's sighting, one of his mates who lived locally saw the creature. It ran out of a woodpile and vanished between some barns. This is interesting as thylacines were thought to make temporary dens that they used for a few days before moving on.

Joe's wife Pat had also seen the Tasmania wolf. Thirty-five years before, in 1981, she had been driving a couple of miles outside of town. It was winter and twilight. A Tasmanian wolf crossed the road in front of her car. She got to within fifteen feet of it. She clearly saw the striped flank and stiff tail. It was eighteen inches to two feet tall with a yellowish-brown coat and powerful-looking jaws. It was somewhat greyhound-like. Pat had it in view for sixty seconds before it moved off into the surrounding fields.

One of Joe's interests was the old convict roads. These were constructed by convicts transported to Tasmania from the 1830s onwards. He and a number of friends try to locate and restore the roads. He took us out to show us a rock that had a bizarre carving on it. It had been made by one Nehemiah Rogers. Originally from Brocking in Essex, and a stonemason by trade, Rogers was born in 1825. Convicted of burglary in 1845, he was transported to Tasmania. Joe didn't know what the strange symbol carved into the boulder represented. He thought it might have been masonic. To me it looked like a stylized, ejaculating phallus.

Joe explained that, the previous year, he and his son had been exploring the wooded hills some miles from the town. His son had been on a gravel path and Joe had been deep in the undergrowth some way from him. Stumbling across some ruined huts, Joe called out to his son. Apparently, his shouting disturbed an animal. His son shouted out to him that a strange animal had emerged from the bush and was on the road a few yards ahead of him. By the time Joe got to the path, the animal had gone. His son described it as the size of a whippet with tan-coloured hair, dark stripes on the sides, and a stiff tail. It trotted off up the track. The creature, apparently a young thylacine, left a set of clear tracks. Joe and his son followed them up the road till they vanished back into the bush. On returning to their car, it seemed that the creature had doubled back and walked around the vehicle before returning to the forest.

Joe returned next day with a camera and photographed the paw prints. They seem to show five visible claw marks on the front foot. The Tasmanian wolf was plantigrade, unlike the placental wolf which is digigrade. This means that it walked on the whole of the foot and not up on the toes like true dogs. The dog's dew claw equates to our thumb or big toe and is held clear of the ground. Clear tracks of a thylacine's front foot generally show five claw marks; a dog's will show four. Also, there was a small indentation behind the metacarpal pad (which equates to the palm) on each print. Again, this is typical of a thylacine.

Mike and I made camp in the area and set up camera traps baited with fresh roadkill or oven-ready chickens. We spent the days exploring on foot and the nights driving.

The following day, we travelled to another town in the area to meet up with veteran thylacine hunter Col Bailey. Col saw a thylacine back in the 1960s on the mainland. He was on a canoe trip in the Coorong Lakes in 1967.

> *"Four hundred yards away I saw a dog-like animal on the water's edge. It was big, like a greyhound, a long animal with short legs, a long tail and a big head. But then it disappeared."*

This fired his interest in the animal, and he moved to Tasmania. Col was lucky enough to meet and interview old bushmen who had been around in the late nineteenth and early twentieth century and mine their wealth of knowledge of the Tasmanian wolf. Without Col's work and diligence, these stories and information would be lost to the ages, as all the old trappers and bushmen have long since passed away.

Col saw the animal again in 1995, this time on Tasmania and in deep bush.

> *"My eyes ran down its back and tail and it hit me—this was clearly a Tasmanian tiger. I was entranced, riveted to the spot. I stood there and watched it for almost a minute before it hissed at me and turned into the bush."*

Beforehand, he had heard the distinctive high-pitched yip of the animal and smelled its pungent odour.

Col kept the sighting under his hat for seventeen years in order to protect the creature.

Col, now seventy-eight, has written three books on the thylacine, *Tiger Tales*, *Shadow of the Thylacine*, and most recently *Lure of the Thylacine*.

We spoke for some time and discussed the power of the animal's bite. A recent paper tried to claim that the animal had weak jaws and would only feed on small creatures like possums. This is totally at odds with contemporary field reports which said the Tasmanian wolf killed and ate kangaroos, wallabies, and full-grown sheep, killing them with exceptionally powerful bites. Several reports said that, when cornered by dogs, a thylacine could bite clean through a dog's skull. A more recent paper refuted the weak jaw hypothesis. Looking at the skull anatomy, its authors concluded that the thylacine had a much more powerful bite than a wolf or dog, but the skull was not as well adapted to hold onto struggling prey. Wolves, being pack hunters, surround their quarry and hang onto it, worrying it to death. The solitary thylacine kills with one or more powerful bites.

Col also spoke of the absurd numbers of sheep kills laid at the thylacine's door in the bounty years. It would have been impossible for the animals to have killed that many sheep without attacking them twenty-four-seven.

Next day we visited a remote valley area before returning to Joe's town. The librarian there had a story to tell. Twenty years before, her car had broken down some miles outside of town. There were no lights, and she was compelled to follow the road in darkness toward the town. She soon became aware of a soft padding behind her. It was too dark to see anything, but she knew that something was following her. She shouted out, and whatever it was ran back into the bush. Later she read of how Tasmanian wolves would often follow men in the bush. She was convinced that it was one such creature that tracked her that night.

We drove down to Hobart to see the thylacine display at the museum. There were stuffed specimens, pelts, skulls, bones, and casts of the last known prints taken in the wild (according to them), but not a word on thylacine survival or the four-thousand-plus sightings since the 1930s.

Later we checked the camera traps. One of the bait carcasses had been partially eaten. A hole was ripped behind the back leg and the internal organs had been devoured. Something had taken the head too. Looking at the pictures we saw two quolls and a Tasmanian devil.

The following day we returned to Joe's to borrow the photographs and make copies. Joe told us of some men out spotlighting who had seen a thylacine just six months before.

He also told us about the shack that was once home to Elias Churchill, a trapper who captured thylacines alive for zoos in the early twentieth century. Col Bailey rediscovered the shanty, not used since the early 1930s, back in 2006. The hut was restored with a grant from Tourism Tasmania. Mike and I decided to take a look at it.

The location was remote and quite a distance away. Following directions and hand-drawn maps, we found ourselves along a track in a wooded area. We failed to locate the hut. Mike walked on ahead and I lingered behind him. On a section of the track, I became aware of a weird smell, somewhat like that of a hyena. I am a former zookeeper and regular zoo visitor, and I am familiar with the smell. The odour seemed to intersect the track and was only in one area. It was as if whatever had left the scent behind had recently crossed the track. The Tasmanian wolf was said to smell very like a hyena.

We tried to find the shack again the next day and this time managed to get to it. The hut was small, and I was impressed that Churchill weathered the harsh Tasmanian winters in the structure. The remains of the stockade where he kept captured thylacines was still standing as well. Churchill snared them. He kept them in the pen, then transported them out of the wilderness on horseback.

We placed a camera trap at the area where the odd smell was detected. On the way back, I found some scat and preserved it in ethanol for analysis. It was dark and fudgy, matching the description of thylacine droppings, as the content of their diet is rich in blood. The droppings that we had found on the last expedition were found to contain bone fragments and were ultimately shown to be from large Tasmanian devils. This sample looked very different, lacking bone chips but containing hair.

Driving down to the southwest, we visited Lake Pedder, Australia's largest freshwater lake. It was once a natural lake of modest size. In 1972, the Hydro Electric Commission of Tasmania flooded the lake by damming the Serpentine and Huron Rivers and extending the lake to its current size of 242 square miles. The project was opposed by conservationists and galvanized the green movement in Tasmania. Tasmanian premier Eric Reece supported the project and gave this appalling quote:

> *"There was a National Park out there, but I can't remember exactly where it was...at least, it wasn't of substantial significance in the scheme of things."*

In 1972, the activist Brenda Hean and pilot Max Price were killed when their Tiger Moth plane crashed. They were flying from Tasmania to Canberra to protest the damming of Lake Pedder; it was alleged that pro-dam campaigners had entered the plane's hangar and placed sugar in one of its fuel tanks.

The flooding led to the extinction of the Lake Pedder earthworm (*Hypolimnus pedderensis*). Another victim was the Pedder galaxias (*Galaxias pedderensis*), a tiny fish found only in the lake. It is now extinct in the area, though nearby populations have been translocated to Lake Oberon in the Western Arthurs mountain range and to a modified water supply dam near Strathgordon.

Sickeningly, big business always seems to triumph over environmental or conservation concerns. Though it looks beautiful, the lake leaves a bad taste in the mouth. There are pressure groups today that are advocating the draining of the man-made lake and the restoration of the original Lake Pedder.

Joe had told us of another local man, Bill Morgan, who had seen a Tasmanian wolf back in the 1970s. We tracked him down and he agreed to talk to us.

A sprightly ninety-three-year-old, Bill had worked for the hydroelectric company. He encountered a thylacine in 1979 but would not reveal the exact location. Bill was in a carful of coworkers. They drove over a bridge and saw a thylacine in the middle of the road. Bill described the animal as "beautiful" with sleek fur and stripes. It moved with stiff-looking hindquarters. It left the road and looked back at them as it went. The group had it in view for six minutes.

His friend Max Macallum also saw a thylacine in the same year. The animal crossed the road in front of him as he was driving to his brother's house.

Bill had recently caught up with his cousin, whom had had not seen in decades. Amazingly, just eighteen months earlier, the cousin and five other people in a car had seen a family group of thylacines. A male, a female, and three pups crossed the road in front of them. It happened on the road to town in West Tasmania. Bill had no doubt that the Tasmanian wolf was still around.

We visited a range of mountains in which the remains of an old osmiridium mining town was located. Osmiridium, a natural alloy of osmium and iridium, was used mainly to make pen nibs. Tasmania was the world's foremost supplier of the alloy. Only a few preserved shanties remain of the town.

We checked the camera traps and found only devil, quoll, and other fairly common creatures in the pictures. The traps were rebaited with fresh meat.

We took time out to visit an artist called David Hurst. He is producing life-sized bronze busts of thylacine heads. He showed us his workshop, where he carved the heads in wax before casting them. They were remarkable in detail. David thinks the animal is still with us and thought that the southwest wilderness might prove a bountiful area.

We visited Joe again, and we headed up to the hills again. He told us of finding the remains of an Aboriginal hearth under a felled tree in the early 1970s. He believed that the hearth had been preserved there for over fifty years.

We met with Kathy Brownie, the proprietor of a local coffee shop. She had played a bit part in the 2011 film *Hunter*, starring Willem Dafoe, Sam Neill, and Frances O'Connor. The film sees Dafoe as a hunter employed by a pharmaceutical company to track down the thylacine. An unimpressive flick, it is filled with scientific errors, such as giving the animal a venomous bite!

Much more interesting was the small museum she maintained in the shop. Among the fossils and minerals were alleged casts of the hind foot of a Tasmanian wolf. They clearly showed the long carpal pad. The casts of the tracks were taken back in 1991 by a guy called Rusty Morley.

Kathy told us that in 1971, she was living in a mining town consisting of wooden shacks and very limited amenities. She said that a bulldozer had uncovered an old thylacine lair that had the remains of prey animals in it. She claimed to have heard the Tasmanian wolf's call on a number of occasions.

Upon returning to his home, we found that seven pigs, escapees from some neighbouring farm, had chomped their way through Joe's potato plants and were now making free with his pumpkin patch. I chased them out of the garden and down the lane with a mop.

As the expedition wound down, we returned to Launceston and visited the natural history museum there. We found it to be better than the one at Hobart. It covered the possibility of thylacine survival and had a map of sightings from the 1930s to the early 1970s. Why this had not been updated was anyone's guess.

The next expedition was provisionally scheduled for February of 2017.

January of 2017 saw my third trip to Tasmania in search of the iconic flesh-eating marsupial known as the thylacine or Tasmanian wolf. I slept for most of the thirty-one-hour journey. Luckily, like Ralph Wiggum, sleep is where I'm a Viking.

I was once again teaming up with Mike Williams from the CFZ's Australian office. We found that having fewer people on the expeditions was easier in terms of organization. We also made less noise in the bush. Again, some people requested anonymity, and we must respect that.

Once more Toyota kindly assisted us by providing a car and a fuel card that gave us unlimited petrol. This took a major financial burden away from us and allowed us to move right across the island.

On the first day, Mike had arranged an interview with a journalist in Launceston. We visited the offices of the *Advocate* and Mike spoke with the journalist involved. I have an inherent mistrust of media types who have misrepresented me and my colleagues in the past. Therefore, I sat the interview out. Mike hoped that a newspaper story would turn up new witnesses and give us leads.

Later, we met up with Granville Batty. Granville was the great-nephew of Wilf Batty, the last man to shoot a wild thylacine on May 6, 1930, at Mawbanna, a hamlet in the northwest of Tasmania. He once believed that his family still had the gun that did the awful deed, but apparently he has since found out that the gun he has is not of a make old enough to have been his uncle's.

Granville is an amiable chap in his early seventies. He has a keen interest in the Tasmanian wolf and thinks the animal may still exist. I had met and talked with him on a pervious trip. Granville had told Mike of a farm where possible thylacine activity had taken place. In the 1990s the place was a sheep farm. The owner had some predator that was taking out sheep on a regular basis. The kills were quite distinct from dog attacks. At the same time, the distinctive thylacine call, a high-pitched triple yap, was heard. One of the workers reported seeing a thylacine from a car in broad daylight. The farmer switched from rearing sheep to rearing cattle and the kills stopped. The calls, however, were still heard by the farmer and his workers.

Then just four years ago, the farmer himself had his own sighting. As with his employee, the animal was seen from a car at the side of the road. It was lying in vegetation, and at his approach it stood up and walked calmly away. He got to within sixteen feet of the thylacine.

We had permission to camp on the farmland, so we drove up and set up some baited trail cams. We made camp, then went for a night drive. As with previous trips, we utilized a crash cam, a windscreen-mounted miniature camera that is constantly recording. We saw Bennett's wallabies, red-bellied pademelons, and a short-beaked echidna. By the time we returned to camp, it had started to rain heavily.

The following day we met up with Granville again. He told us that, whilst trapping possums in 1963 in the area, he had accidentally trapped a huge tiger quoll or spotted quoll. These are flesh-eating marsupials related to the Tasmanian wolf. They have dog-like faces and bodies not unlike a cat's, but somewhat more robust. They are a fawn colour with distinctive cream-coloured spots. The tiger quoll is generally the size of a large domestic cat at around seven pounds. The creature Granville had caught was much larger, comparable in size to an Australian cattle dog, a breed that weighs thirty-three to forty-nine pounds (similar to a border collie). Granville indicated the height of the animal with his hands at around eighteen inches.

This is not the first account of a giant-sized tiger quoll we have come across. On my first trip to Tasmania, we interviewed a man who ran a market garden. We were mainly talking about a thylacine sighting he had in the 1990s, but he also told us that back in the 1960s, he had shot a huge quoll. The animal had been killing his chickens. He said it was the size of a cattle dog with a thick bull neck. He raised his hand to indicate the length of the creature when hung up. It was around five feet long. We heard of another of similar size seen in the Cradle Mountain area.

Granville wanted to give the animal to a local zoo, but was too scared to tackle the angry beast. He was only fifteen at the time. He told the farmer whose land it was on and intimated that he should give it to the zoo. In an act of wanton cruelty, the farmer threw the quoll into a barn with four dogs that ripped it to shreds. Apparently, the dogs were also badly injured by the giant quoll's teeth and claws.

It is possible that these outsized quolls represent a new species or, more likely, freakishly large individuals of the tiger quoll. The Queensland tiger is a cat-like marsupial the size of a puma reported from the tropical north of Australia. Some cryptozoologists have postulated that it could be a surviving form of *Thylacoleo carnifex*, a long extinct flesh-eating marsupial distantly related to the wombat. Perhaps some huge form of quoll is a more likely candidate if such a creature exists.

Granville had also mentioned that he had a "shack" in the area that we could use. Despite the fact that he told us that it had no electricity, we jumped at the chance. Mike had been sleeping in the car and I in a tiny, one-man tent. I loathe camping, despite having to do it at length on almost every expedition I've been on. I seldom get a good night's sleep under canvas, finding it cramped, cold, and uncomfortable.

Granville took us to the aforementioned shack. Mike and I were expecting some malodorous shanty one step from a garden shed. In fact, the "shack" was a four-bedroomed farmhouse with a kitchen, living room, shower, and toilets. The only reason it had no electricity was that Granville had not yet turned it on. One flick and we had light and heating. We used the farm as a base whilst we stayed in the north.

After settling in we did a night drive, seeing wallabies, possums, and Tasmanian devils.

The one minor downside of our Toyota sponsorship was that the fuel for the car could only be bought from one specific company. All the garages in towns nearby had changed since our last trip, and now the nearest garage of the company in question was many miles away in Wynyard. We brought some large fuel containers to stock up and avoid wasting time on too many long runs for petrol.

In a small museum in Wynyard, we came across a pamphlet, *The Tasmanian Tiger Trail* by Colin Berry. It contained accounts of a number of sightings, some of which I had never heard. The publication had once come with a DVD that apparently contained an interview with the late Wilf Batty. The museum didn't have a copy, but Mike wanted to try and track it down.

During a night drive, we took a wrong turn on the overgrown, labyrinthine tracks in the wilderness. We spent the better part of an hour lost and had to stop the car several times to pull logs, fallen trees, and branches out of our way. We finally found the right path and made our way back.

Next day we returned to Wynyard for a meeting with a councillor who was, many years ago, involved in the publication of the thylacine pamphlet and the accompanying DVD. He could not find the disc in his archives and was sure that it was actually just audio rather than film.

We met up with a man who had seen a thylacine on the farmland we had camped on. In 1990, he had been driving along a wooded road on the property with two passengers in his car. All of them saw a Tasmanian wolf emerge from the vegetation at the side of the road. It seemed like it had been resting there and got up when it heard the car approach. The animal looked at them before walking off into the forest. All of them had a clear and good view of the animal in broad daylight.

Mike's interview had begun to attract callers with their own stories. One man, who claimed to be an experienced hunter, said he saw a thylacine in Queensland on mainland Australia many years ago. He said that he and some friends saw it emerge from some bushes. His description of the animal, however, lacking stripes and having a moth-eaten look, was that of a dingo with mange, not a Tasmanian wolf.

We collected our camera traps. On the way to get one of them I heard a high-pitched yip. I froze in my tracks, straining to hear. Nothing came. I moved on again and once more the yip sounded. It was a single yip, not the triple yip associated with the thylacine, but it still intrigued me. As I moved once more, the yip was heard again, then I realized it was my boots squeaking! We checked the cameras. They showed Tasmanian devils, wallabies, and quolls.

We drove down to the tiny village of Corinna. The place is named after the Tasmanian Aboriginal name for the thylacine. The drive was a long one through very wild territory on poor roads. The village itself is tiny, consisting of a small pub/restaurant and a handful of houses. It is beside the Pieman River and was a former mining colony. Unfortunately, the ferry across the river was broken and we could go no further.

We moved on to Derwent Bridge and Mike had another contact, this one more promising than the last. The sighting had occurred when the witness was a boy in 1951. It occurred in the Central Highlands at five in the morning when the witness was travelling with his father, a farmer. They saw a pair of eyes on the dark road ahead and thought it was a calf that had wandered out onto the road. As they drew alongside the creature, the boy looked at it from a distance of only three feet from the car's window. It was a dog-like animal with striped hindquarters and a long, stiff-looking tail.

The boy was later interviewed by Dr. Eric Guiler, a zoologist from the University of Tasmania. Guiler dedicated much of his life to researching the Tasmanian wolf. Guiler apparently believed that the boy and his father had indeed seen a thylacine.

Later we moved down to Derwent Bridge on the edge of the Franklin Gordon National Park.

A man called Roy told us of a man he knew who had an odd encounter in Wuthering Heights, a coastal plain near the Frankland River in northwest Tasmania. Some years ago, the witness, a logger, had been in camp with several friends. They heard a yip-yip-yip vocalization and a crashing in the undergrowth. Suddenly a wallaby exploded from the undergrowth and ran toward the men. The animal seemed exhausted and was panting. It actually hid behind the informant's legs. It was obviously very scared, but the men never saw what was chasing it. He thought that it was being hunted by a thylacine.

More night drives revealed lots of wallabies, but no devils or quolls. It seemed that these predators were lacking from this particular area.

We met up with Col Bailey, who I had been introduced to on the last trip. Col is unquestionably the greatest living thylacine hunter, and as well as stalking the beast through the wilderness (an endeavour that rewarded him with a sighting in the 1990s), he also interviewed old bushmen, loggers, prospectors, and hunters back in the '50s and '60s. These people had first-hand experience of thylacines in the Tasmanian wilderness during their official period of existence. They are now all dead, and without Col's efforts, their knowledge and stories would be lost to the ages.

Col told us about his expedition into the far southwest of Tasmania. The Southwest Conservation Area is an uninhabited section of the island. A wilderness of windswept button grass, it is seldom trodden by man, and has only one rough track leading into it from an arm of the Macquarie Harbour, a large, shallow natural inlet. The track peters out after a short while.

Col originally planned to go to the area by boat, but the swell was too great in the Southern Ocean. He ended up chartering a helicopter to drop him on the west coast of the area, an endeavour that cost him five thousand dollars and today would be far costlier. Col spent a week at the ends of the earth, trekking up and down the coast and venturing inland. He found possible tracks on a remote beach. He almost ran out of water and found that the creek near his camp was salty.

We stayed with Mike's friends, a couple who ran a little goat farm near Bronte Lagoon. They had met a man in 1967 who was driving from Queenstown to Hobart. At one point, he said he had seen a dog-like animal with stripes. He had never heard of the Tasmanian wolf and had no idea of the importance of his sighting. The couple's daughter had also seen a thylacine in 1985. Whilst walking home from Deloraine, she saw the creature in a meadow.

One of their friends had a strange encounter in 2006, but one not related to the Tasmanian wolf. The woman had been kayaking on the Mersey River. Some huge aquatic animal swam up to her kayak. Whatever it was, the thing was pulling a large wake and disturbed the witness. He thought it could possibly be a giant eel.

The following day we met up with Lloyd and Maureen Poke. The couple used to have a farm in the northeast of Tasmania. When they were there from the '70s to the early '90s, it was still a wild place. Now the area has been swallowed up by farms that have sprung up all around. During their years on the farm, they both claimed multiple sightings of the Tasmanian wolf.

Lloyd had his first encounter with the creatures as a boy in 1957 near the Ouse River. He watched a family of thylacines, a mother with three cubs, walking through the scrub. He decided not to tell anyone.

In 1986, the couple were driving along a road on their property when they saw a strange beast in the road ahead. Maureen said, "I realized it had stripes and could not be a tiger quoll. It had to be a Tasmanian tiger."

The creature was around 164 feet from the witnesses and had a striped back and stiff tail, a description that should be familiar by now!

The following year, Lloyd noticed that something was taking dead wallabies that he had shot on the property. He was using them for dog food. Intrigued, he brought one thousand feet of tough cotton and tied it to a wallaby carcass. Later, when the carcass had been taken, he followed the cotton, like Theseus following his ball of twine in the labyrinth of the minotaur. The cotton led him through the forest to a lair under an old tree stump. At the time, he had no camera, so he set up a tape recorder. He was rewarded by recordings of crunching noises and odd screams. Lloyd suspected that a Tasmanian wolf was responsible.

As it turned out, he was correct. Sometime later he saw a thylacine chase a wallaby into the scrub. He found it biting into the wallaby's chest. The predator stood up on its hind legs and gaped its formidable jaws in a classic thylacine threat display. Lloyd backed off immediately and left the animal to its meal.

Another time, he was out with his dog and crawled into a wallaby trail in the bushes. He heard a snarling from further along the trail and retreated.

Lloyd's closest encounter happened in 1990, after he had put in a new fence. He saw a Tasmanian wolf walking along the fence and managed to corner it. He tried to catch it by grabbing it. "I grabbed at its neck thinking that there would be loose skin, but it was tight and muscular. I couldn't hold onto it and it struggled free." Turning around, it kicked out, scratching Lloyd's arms, then leapt over the fence and disappeared.

The couple left the farm in the mid-1990s.

Each time I have returned to Tasmania, I have had my conviction that the thylacine is still alive and well reinforced. I have not revealed everything that happened on this last expedition. Suffice to say, I have seen some convincing evidence about which I have been asked to keep quiet for the time being. I feel it is now just a matter of time before the Tasmanian wolf is officially removed from the list of extinct animals.

The Almasty

I first heard of the Ukrainian biologist Grigory Panchenko in Dimitry Baynov's book *In the Footsteps of the Russian Snowman*. It detailed a close encounter he had had in a barn with the almasty, a relic hominid said to inhabit much of Central Asia and the former USSR. Some years later, I read about his long-term work in the Caucasus Mountains and the fact there were many reports from the Kabardino-Balkaria part of the range. Panchenko believed that the population of almasty was increasing in the area.

The almasty, variously known as almas or albasy, is said to be a man-like, hair-covered beast. It is smaller and more human in its appearance than the yeti or sasquatch, but larger and more muscular than a man. It is generally thought to be of the genus Homo rather than a pongid. Records of it go back hundreds of years in Central Asia, and it was included it catalogues of local wildlife.

I thought it might be a good idea to contact Grigory and try to do a joint expedition with CFZ members and his own team. Getting hold of a number and email for him took some time, and it was the best part of a year before I was able to contact him.

We invited Grigory over to speak on the cryptozoology of the Caucasus at 2007 Weird Weekend. Grigory had a vast amount of information, most of which had never been aired in the West. As well as hominids, giant black snakes up to thirty-three feet long had, for centuries, been reported from the area. His talk proved to be one of the highlights of the conference. Between us, we arranged an expedition for June/July of 2008. Grigory and his colleagues would be in the field for two weeks prior to the CFZ team's arrival.

The British contingent, apart from myself, consisted of Dr. Chris Clark, stalwart of most of the CFZ's previous expeditions; Dave Archer, a CFZ member who has organized his own expeditions in the past; Adam Davis, an experienced traveller and cryptozoologist of Extreme Expeditions; and Keith Townley, a friend of Adam's who had accompanied him on some of his past adventures.

Shortly before we left, we were contacted by Professor Bryan Sykes of Oxford University. Professor Sykes is one of the world's leading geneticists. He was interested in possible hominid survival and possible interbreeding between hominids and modern humans in past centuries. During his seminal work in mapping the human genome, he was unable to find any DNA traces of anything other than modern man. He offered to do analysis on any samples we brought back, and wanted us to carry out swab tests on Balkarian people.

After a mind-bending ten-hour wait in Moscow, we flew to Mineranye Vody, where we were met by Grigory and Alexey Ahokhov, a very tall Russian computer expert and archaeologist. Alexey had a delightful dog called Humma, a cross between a red setter and a spaniel. She accompanied us on all our Russian adventures.

After a night's stay in a spartan hotel in Tyranyauz, we drove up increasingly poor roads near crumbling cliffs and riverbanks to the area of our first investigation, White Rock. We were introduced to Anatoly, the final member of the team. He was a Ukrainian archaeologist with a ginger beard, little English, a great sense of humour, and an even greater love of vodka.

We made our camp in a small valley. When the road was first cut into this area of the Caucasus in 2000, the workmen sliced through many ancient tombs on the way. Around a thousand tombs are scattered around the area. Many, bisected by roads, now spill their contents to the floor. Dozens of human bones and skulls were just sticking out of banks all around us. The remains were of Sarmatian people, who originated in north Iran. The nobles were buried in cliff faces and slaves in the lower areas. The tombs dated from the third to the seventh centuries. The slave's skulls had an odd domed appearance due to ritual binding. Grigory said that when he first saw one, he thought it was an almasty skull, but soon realized that the bone was not thick enough.

In the two weeks before our arrival, however, Grigory had uncovered what seemed to be skull fragments of an inhuman thickness in a cave in the cliffs of White Rock. On examining these, I agreed that they did indeed seem too thick to be from a modern human. We bagged them up for analysis in the UK. We also took some of the Sarmatian bones, in case their DNA had any odd markers that might hint at hybridization with relic hominids.

White Rock itself rose cloud-festooned and sheer above our camp. Behind it was a range of ragged mountains called "The Stepmother's Teeth." The area was home to a she-bear and her two cubs, but we were not lucky enough to see them.

Anatoly told us of his own encounter with an almasty in the 1980s. He was staking out an abandoned farmhouse near Neutrino. From a hiding place, he saw a specimen pass by only thirteen feet away. It was six feet tall, but powerfully built. It had grey hair "the colour of a poplar tree's bark." Its head was domed with a sagittal crest and its nose was human-like, but smaller. It had no chin, and a thick, short neck. It swung its long arms as it walked.

Anatoly had also seen one of the giant snakes in a cave in the south of Kabardino-Balkaria, near a town called Sammakovo. He was being lowered into the cave when he saw a black snake he estimated to be twenty-three feet long, swimming away in the water that filled the cave. His father had also seen such a monster snake many years before in Kazakhstan. Whilst in a marsh he saw what, at first, he thought was a man. As he drew closer, he thought the tall, dark object was part of a dead tree. The he realized that it was a huge snake rearing up like a cobra.

The following day, we set out to investigate the cave where Grigory had found the skull shards. Confronted with the massive climb, Keith decided he could not make it and turned back to camp. The rest of us began on the long and winding path upwards. Eventually we left the path and climbed the increasingly steep slopes through flower-strewn alpine meadows. We paused occasionally to catch our breath, and eventually we reached the cliff face.

We walked along a narrow path to the cave were Grigory had found the skull fragments. An excavation of the cave revealed no further bone remains were found, but some interesting dung samples were. We spent the day exploring more shallow caves and then set up two camera traps. Dave tried to climb up the near sheer walls of one of the cliffs, but the rising winds finally made him turn back.

The following day we climbed again (sans Keith) to retrieve the cameras. Plugging them into Alexey's laptop, they showed only the setting and rising of the sun. On the way back down, Alexi saw a wild cat.

That night our camp was disturbed by the unearthly cries of a jackal. It came right through the camp in the wee small hours. We scrambled out of our sleeping bags in an attempt to photograph it, but by the time we were up, it had vanished into the night.

The next day we split up. Keith, Anatoly, and I investigated a wooded area whilst Dave, Grigory, and Adam went off to interview a man named Surgit who claimed to know the whereabouts of an almasty's body.

On the way to the forest, we investigated some small caves. In one, Anatoly had found eleven human skeletons dating back three thousand years. No evidence of the almasty was found there, just some badger droppings.

From a distance it looked like the woods were surrounded by grass. In fact, it was a carpet of six-foot nettles. The going was steep and slippery. At one point, Anatoly pointed out scratch marks on a tree. They looked to be made by nails rather than claws. Close by, Anatoly found a hair in a tree. It was long, stiff, and bi-coloured. At first, I became excited, but then I found further clumps of hairs, and they began to look suspiciously like the bristles of wild boar. I bagged them anyway just to be sure.

The other team, headed by Adam, who is a presenting officer for the Home Office, interviewed Surgit. Adam's job basically means cross-examining people who want to stay in the UK to see if they have any case. Hence, he is an expert at interviewing. Surgit claimed to have found the body of a female almasty in 1996. It had been crushed under a rockfall on a mountain known as Kashkatash. He had retrieved a tooth, which he had given to a friend. Surgit said he could lead us to the place in question. Adam seemed convinced he was telling the truth.

Anatoly told us some interesting cryptozoological snippets. Some years ago, a friend of his was on a boat in the Lena River in Siberia when he encountered a strange creature. It had a black, humped back and a six-foot-tall fin. It reminded him of a killer whale, but they were thousands of miles inland at the time.

A man on the boat took two shots at the beast with a rifle. It turned and swam at speed toward the boat. The man pumped three more bullets into the creature, and it dived under the boat and swam away. The description recalls creatures described from Lake Vorota in Siberia. The beasts here are up to thirty-three feet long, and have a dorsal fin and a wide head. Could they be some form of colossal fish?

Another river where he was told of a monster was the Don River that runs into the Black Sea. Here he was told of a twenty-foot wels catfish that had overturned a boat and eaten a man and a woman. The wels catfish is a contender for the largest freshwater fish in in the world (along with the Chinese paddlefish). The largest known specimen was sixteen feet long. I have seen an eight-foot one and that was spectacular enough to be called a "monster."

Grigory too had some tales to tell, and not all of them about monsters. Grigiory told us why he avoids vodka and only drinks a little beer. Whilst in national service in the Ukraine in 1986, he was involved in a fight. The older soldiers attacked the new interns, and Grigory was punched in the temple with a knuckle duster. He woke up in hospital, only to find that the rest of his platoon, including the man who attacked him, had been sent to Chernobyl! The plant had gone off pop whilst Grigory had been out cold in a hospital bed!

His files on the almasty were extensive, and he shared some of the more unusual stories with us. One story involved a farmer whose savage Caucasus shepherd dogs were going wild. On opening the door of his house, he was alarmed to find a young almasty apparently trying to escape the dogs. It punched the man in the shoulder and knocked him down. The creature ran away, pursued by the dogs. The hounds later returned with blood on their fangs.

On another occasion, an adult almasty approached a house and was attacked by a big dog. The almasty used a club to bludgeon the dog to death. It then entered the house and stole a large Balkarian cheese.

The strength of the almasty far, far exceeds that of any modern human. On one occasion, one was observed fighting a bear. The almasty punched the bear, which tumbled over, then retreated. Grigory thought it was a young bear, as an adult male would be more than a match even for an almasty. Indeed, almasty hair has been found in bear droppings in the Pamir Mountains. He has also been told of almasty remains from a specimen killed by wolves.

One man saw an almasty close to his house and, worried about it stealing food, he threw a stone at it. The almasty retreated behind the house, and soon after, a huge rock was hurled right over the house, narrowly missing the man. In the morning, it took two large men to lift it.

Another man struck an almasty that had entered his house. The creature hit him back and knocked him fully fifteen feet.

Grigory also noted that the almasty seemed to be left-handed.

Anatoly told us that there were, in fact, two man-like creatures in the Caucasus. The almasty, which is the smaller and more human-looking of the two, and the much bigger, more ape-like mazeri. The mazeri more closely resembles the larger type of yeti and the sasquatch. It stays away from humans, whereas the almasty will approach humans and human habitation.

The next day we decamped and headed for the Elbrus area, where the body was supposed to be. This was near the border with Georgia, so we had to get border passes. We stopped in a campsite for mountaineers that consisted of a number of alpine shacks of Moominesque cuteness. Surgit arrived and introduced himself. He looked a little like Father Ted. He, via the translation of his beautiful daughter Tanya, explained that the area where he saw the body was only half an hour's walk up a nearby mountain. He was dressed in slip-on shoes and a shirt. At only half an hour's walk, the body must have been in the foothills and easily accessible. We all set off immediately, apart from Keith who stayed in the wooden shack.

The going was a little steep, but after half an hour on a decent path, we reached an area of scree and snow that I assumed was where the body lay. To my surprise, we passed this area and carried on climbing. We passed a huge boulder that had about a dozen plaques on it. They were memorials to people who had died on Kashkatash! We had expected an easy thirty-minute jaunt up a hill. Instead, we were climbing to the top of a mountain that had killed a considerable number of people, whilst dressed in a fashion more in keeping with a walk in the park!

To one side of the mountain was a retreating glacier that had carved out treacherous cliffs. Masses of loose scree coated the sides of the mountain and great sheets of frozen snow, as slippery as a politician, stretched out over huge areas.

The path disintegrated, and the way grew ever more treacherous and steeper. In such situations a group should stick together, but that is exactly what didn't happen. Some forged on ahead whilst others lagged behind. We were led across a field of ice and snow that sat atop jagged boulders. Chris managed to walk across it, but I am twice his weight. My leg went through the ice, and I lost balance. In a flash I was sliding on my belly down the steep ice sheet toward spiky boulders three hundred feet below. Realizing that cryptozoologist puree was soon to be on the menu, I swung my body around and jammed my boots onto the rocky outcrops at the side of the scree pile I had been walking on prior to venturing onto the ice. This stopped me after about thirty feet, instead of the three hundred that would have ended in a sticky splat.

I crawled back onto the rocky area and decided that the ice was impassable for me. I stayed put whilst the others carried on to the area with the body. We had not started until four in the afternoon and now the sun was getting dangerously low. Being stuck on the mountain after nightfall would be as dangerous as French-kissing an industrial meat-grinding machine, so I began to get concerned. The others returned after a cursory look at the place in question. Then we began our return journey.

On the way down we got split up again. Taking a wrong turn, I found myself walking along a wasp-waisted path next to fifty-foot sheer cliffs constructed of a particularly loose and crumbly soil. At one point this gave way beneath me, and I had to grab onto tree branches and hang, Indiana-Jones-style, over the cliff. I managed to pull myself back up and continued gingerly on my way. I was glad to finally get to the bottom, having nearly died twice in one day!

Back at the camp, we had beers and shish kebabs and talked about the need to stick together. The next day we returned to Mount Doom (except for Keith, who had more sense) and climbed up a route that avoided the ice. Why Surgit had led us up the ice-festooned way the day before, no one knew. This route took us higher, and we climbed up through swathes of dwarf rhododendron.

At a height of nearly 10,000 feet, I began to suffer from altitude sickness. On a steep area, my vision failed like a camera iris closing. Blackouts at 10,000 feet are not good, so I decided to stay put. I sat down to try to get myself together as the others carried on higher. After about two hours, Chris and Adam returned. The area was so steep and cramped that not everyone could work on it safely. They left Dave there with the guides and Surgit, as he dug holes for a living. The three of us wearily returned to the camp.

Some time later, Dave and the others returned. Dave was holding a large day-glow orange body bag. He had found a high-altitude cave with a nest made of rhododendrons inside. He had stuffed the nest into the bag and brought it down. Donning plastic gloves Adam, Chris and I began a careful sifting of the vegetation. We found and bagged over twenty hairs with medullas as well as two pieces of dung.

The following day, we headed for Elbrus village (except for Dave who, like a glutton for punishment, had decided to go up the mountain for a third time). We were to interview an elderly man who told Grigory that he had seen an almasty in his youth. However, when we interviewed the man, who was eighty-five, his story had changed. He said that it had been his father who had seen the almasty at the age of fourteen. This would have put the event in the 1890s!

The old man recounted what his father had told him. It had been around noon, and he had opened a door into a room in a part of the house where the celling had collapsed. He saw a young almasty sitting in a chair. It seemed to be basking in the rays of the sun that fell through the roof. It was covered with hair. The hair on the face was reddish. It had long, fine hair on its head. The eyes were red, but the old man thought his father had meant red-veined rather than glowing red. The creature threw its head forward and the long hair fell in front of its face. The witness quickly shut the door and retreated.

The old man also said his father had seen a big snake near the house the man currently lives in. It was in 1964. His horse had reared up and he saw a grey-green snake thirteen feet long and as thick as his arm, slithering away.

Later we talked to a man of about thirty named Tahir, who was the vice president of Elbrus National Park and a doctor of Geographical Science. He told us that three years before, whilst hunting for some lost sheep, he had encountered a big almasty. He had been walking through a sparsely wooded area at twilight when he saw what he thought was a cow lying down. Then the "cow" stood up, revealing itself to be a tall, man-like figure. Thinking it was a human (the figure was in silhouette), he asked in Balkarian if he had seen any sheep pass by. When no answer was forthcoming, he asked the same question in Russian. Still there was no answer. As he drew closer, he saw that it possessed a high, dome-shaped skull. Then he realized that it was an almasty. He decided to fetch his uncle to show him the creature. Looking back, he saw the almasty walking off into the hills. By the time he returned with his uncle, it had gone.

Our next port of call was the small town of Neutrino. We had rented a small, spartan flat in a tower block. Eight adult men in a tiny flat was a squeeze. There was rarely any hot water, and the electricity supply was not very reliable either. I have never been anywhere as depressing as Neutrino. It is not mean or dangerous like Georgetown or full of beggars like Banjul, but it has an air of decay and hopelessness quite unlike anywhere else I have been to. The tower blocks are crumbling and peeling. Many stand empty, and some were never finished before the collapse of the Soviet Union brought economic degradation to the area. There is a 90 percent unemployment rate. The bleak, Eastern Bloc architecture is totally at odds with the lovely-looking mountains that rise up as a backdrop to the town.

Adam and Dave went with Anatoly to stake out an abandoned farmhouse about a mile out of town. The house had a weird history. In the early 1970s, it was supposedly the scene of a triple murder. An old man had some money put away. He had decided to spend it. Three of his relatives got wind of this and went around to his house to try to force him to hand over the cash. There was a struggle, and the old man was killed. His wife stabbed the killer to death, but was then killed by the remaining two brigands. They ran into the mountains, but were later found by the police. The farm has stood empty ever since.

It was the building where Anatoly had his almasty sighting back in the 1980s. In 2005, it was the scene of a very close almasty encounter. Three shepherds had been using it to have a drink in. The door to the veranda opened and a big male almasty walked in. It picked up the nearest shepherd and gently put him to one side before leaping off the veranda.

Grigory, Chris, Keith, and I went to investigate some caves were Grigory had uncovered some human-like bones but left them in situ. The caves had filled with earth over the past few centuries and were now little more than crevices. Grigory crawled in with a trowel and started excavating. There were two collections of bones, but both were clearly human. One was an old woman with only one tooth left in her lower jaw. A coccyx and some ribs from this individual were also found. Grigory reckoned that they dated back about two hundred years. Upper and lower leg bones from a man of an earlier age were also found. All were packaged up for testing in case they had any odd markers in their DNA that might suggest hybridization with almasty in past generations.

In the early evening, back at the flat, Keith was on the porch, looking down at the street below, when he beheld an amazing sight.

"There's a cow here, eating a fire," he said. It was true. A brown cow was munching happily away at a small bonfire lit on some waste ground below the flats. Another cow tried to join in the igneous feast but was jealously chased off by the first cow, who defended her flaming snack ardently. At one point the cow withdrew momentarily as if its lips were scorched. It soon resumed fire scoffing. It was obviously enjoying its meal, as it was drooling prolifically! Perhaps the cow was trying to eat the ash and charcoal for its mineral content.

In conversation Grigory revealed two more priceless cryptozoological gems. Whilst researching in a Moscow library some years ago, he came upon an amazing story, complete with a photograph, in a 1928 edition of a magazine called *Knowledge Is Strength*. The story told of two strange creatures encountered on the coast of the Barents Sea by hunters. They were otter-like in shape and bounded across the land and into the sea. The hunters managed to shoot one of the large creatures but the other escaped. It was, apparently, very hard to kill.

The photograph showed the hunters with the skull and hide of the creature. The skin looked like that of an elephant seal but hairy. The skull was thickset and had eight interlocking teeth. The hunters sent details of the creature to Moscow University and received a letter saying, "Thank you for this information. No such animal exists."

Another story related by Grigory involved a friend of his who had been a geologist specializing in searching for oil deposits. In the early 1980s, he had been in eastern Siberia carrying out investigations into possible deposits. He was working north of Vladivostok when some locals informed him that something like a "dead crocodile" had washed up on the beach. He took a look at the carcass and photographed it several times. It was a large, elongated creature with only four teeth. These were arranged two on the upper and two on the lower jaw. The teeth were at the very tip of the jaw. He did not know the significance of the find and it was soon washed back out to sea.

Sadly, after the breakup of the Soviet Union, he lost his job and subsequently his house and family. He ended up drinking himself to death. Grigory, however, had copies of the photographs and thought that they might be the remains of a primitive whale or archaeocete. However, when he sent copies to me, it was obvious the creature in the pictures was nothing more than a dead beaked whale.

At the farm, Anatoly said he had heard a male almasty vocalizing to attract a mate. Adam had heard some weird crashing noises, but no one had seen anything. The camera traps they had set up around the farm and its outbuildings revealed nothing but branches and grass moved by the wind.

Next day I resolved to join them on a second vigil at the farm. During the day, Grigory, Chris, Dave, and I set out to climb a mountain called Gobisanti to investigate an avalanche. Avalanches kill mountain animals such as wild goat and yak that the almasty will then feed on.

As we set out to the foothills, the air was split by a loud, inhuman-sounding bellowing emanating from behind some bushes. Grigory immediately said that it was no animal he knew of. The harsh noise continued and got louder. Chris, Dave, and I fanned out around the bushes in a pincer movement, cameras at the ready. Could we have disturbed a sleeping almasty? As we drew closer, something loomed from the bushes. It was Alexey and Humma, who had taken a short cut ahead of us and then hid and made some spectacularly inhuman noises to scare us!

Grigory had warned us that we might have to cross some streams on the way up Gobisanti. These "streams" were in fact increasingly dangerous rapids that we had to cross on foot. On either side, we had to navigate endless legions of rocks and boulders. It was slow, tedious, and exhausting. We finally reached the avalanche area. The snow had retreated and frozen over, but hundreds of pulped trees lay strewn around like matchsticks. The only dead animal was a rancid cow too foul for an almasty to eat.

After a quick lunch below the snowline, we headed back down. The rapids were getting ever fiercer and crossing them harder. One area was particularly savage. We got a large log and braced it over the rapids. They were not wide but very fast. With Grigory holding one end and Dave the other, I tried to haul myself across. Halfway, the pull of the cascading water sucked me down and my grip faltered. Grigory wrenched me out and onto the bank, saving me from being smashed into the rocks by the thundering cascade.

Next Chris tried his luck. He came even closer to me than death and was plucked out by Grigory. We finally staggered back to Neutrino damp and shaken.

That night Anatoly, Dave, Adam, and I did a stakeout at the abandoned farm. The building consisted of three rooms, two of which were locked. Around this in an L shape ran a veranda with a door at one end. The main building was surrounded by other smaller outbuildings.

We set up camera traps in four locations around the grounds of the farm. Anatoly brewed up red wine and honey on an old stove in the hope that the smell would attract the creature. We also laid out bread and honey.

We all took up posts in various places on the veranda as night fell. The hours seemed to go quickly as I sat staring out into the darkness, listening for the slightest sound. Around ten thirty at night, something made a bird-like twittering noise. Shortly afterwards, one of the camera traps fired. The almasty is said to make a twittering sound. One of the specimens Grigory saw was making such a noise. Anatoly went out to investigate and did not return.

Dave fell asleep on one of the manky beds in the open room. Adam and I sat on the other one, listening intently. A lull in activity was supposed to occur around midnight to three o'clock in the morning. Hence Adam and I had entered the room to warm ourselves around an old stove. The seven-foot door of the room was open an inch or two, and starlight from the clear night was pouring in. At around two thirty in the morning, Adam and I heard a deep, guttural vocalization. The nearest phonetically that I can write this is "bub-ub-bub-bub."

"Did you hear that?" I whispered?

Adam nodded solemnly.

Shortly after, something passed by the door, blocking out the light momentarily. Whatever it was, it was large enough to put the seven-foot door in shade, and it seemed to be walking along the veranda.

"Did you see that?" I asked.

"Something is on the veranda," said Adam.

Adam and I grabbed our digital cameras and rushed out, to find only darkness and silence. We did a circuit of the building with our torches but found nothing. Did an almasty pass by us only twelve feet away on the veranda? I don't know. If it did, it was as fast and silent as a cat. But something blocked out a slit of starlight seven feet tall only seconds after the weird vocalization.

At first light, we looked for Anatoly. We were worried that he might have fallen in the dark and hurt himself, or even been attacked by a bear. We found him asleep in one of the outbuildings. We took the camera traps back to the flat and downloaded the images onto Alexey's laptop. They showed sunrise, sunset, and branches moved by wind.

The twelfth day was to be the last for Adam, Keith, and Dave. Surgit phoned, telling us that the friend he had given the almasty tooth to was a wise woman who was using it as a charm. She worked in a restaurant in a nearby village and he would take us to her. He duly arrived and we set off.

The wise woman was not the crone that the phrase "wise woman" conjures up, but a fairly normal-looking middle-aged woman. We had a pleasant meal in the restaurant whilst Surgit spoke with her. Apparently, she had given the tooth to her daughters in the city of Nalchik. They had mislaid it and were now tearing apart their flat in order to find it. I found it very odd that an object used as a "charm" was being treated so offhandedly.

That night we all drank beer, wine, and vodka to see the three lads off. Alexey drove them back to Mineranye Vody and picked up his girlfriend Natasha.

Next day there was still no sign of the elusive tooth. Chris, Grigory, and I staked out the farm again, this time adding pungent-smelling fried onion to the bait. Nothing happened, but I had recurring nightmares. Each time I fell asleep, I dreamed that something with long, bluish fingers and dressed in a blue-black, monk-like robe was trying to strangle me. They were lucid dreams in which I struggled to wake up.

Back at the flat in the morning, Surgit arrived and told a strange story. He said that the spirit of the almasty had made the tooth vanish as well as the body on the mountain. However, he announced proudly, djinn (Islamic spirits) had given him a red hair from the queen of the almasty. He presented us with this item. It looked more like a vegetable fibre to me.

What were we to make of this wild tale? Grigory was rolling his eyes whilst translating. Surgit never once asked for money. He also trekked up the mountain again and again and worked hard looking for the supposed almasty corpse. The only conclusion we could come to was that he believed his own story and must have had some kind of mental problem. The promising lead of a corpse and a tooth vanished in a puff of smoke.

We phoned a man called Saeh Kumbunov whose number Surgit had given us. He said the man was with him when he found the body. The seventy-year-old was much surprised at Surgit's claim and said there had never been a body in the first place.

Later that day, a gaggle of local women burst into the flat jabbering excitedly in Balkarian. They looked like stereotyped peasants in some old film. It turned out that the person we were renting the flat from was not its owner. It looked as if we were going to be ejected onto the street! Grigory seemed to pacify them after a bit and they relented and let us stay. After that, the electricity went out.

Alexey returned with Natasha, the editor of a furniture magazine. We were also joined by two Russian bodybuilders called Sasha and Victor. We departed for an area called Gushgit. We drove as far as we could, then walked up a long, winding steep path into the hills. We made camp, then went off to explore a kosh or shepherd's house. The almasty is often said to lurk around these, as some shepherds put out food for them. This one was a long-abandoned and malodorous shanty that no self-respecting relic hominid would be seen dead in.

In the morning we set off to explore a series of shallow caves in the high mountains. We found some hair and a lot of dung. Grigory also unearthed what may have been finger bones. We carefully bagged all the material. We came upon another abandoned kosh. This was in a worse state than the first, but it had an interesting Balkarian tribal symbol on the wall.

The following day, I felt totally drained and had to turn back. I got to camp and collapsed into the tent. I made the right choice. A violent thunderstorm struck, and the others returned soon after, apart from Anatoly, who had vanished once again.

In the morning Anatoly returned. He had found a cheese factory deep in the mountains! There were guards with dogs and machine guns that were often fired randomly into the night. He thought that any almasty would have long left the area. We returned to Neutrino.

We visited Elbrus village again on the track of eyewitnesses. Grigory was hoping to track down the shepherd who had been lifted up by the almasty at the old farm in 2005. We found out that this man was away at a funeral and wake for several days, but we did find some other witnesses, who we interviewed.

One old man called Bahua Tilov had seen almasty on several occasions since the 1970s. The first time was whilst he was working in irrigation near Neutrino. He saw a large black almasty with two smaller grey-coloured ones sitting amongst the rocks. As he approached, the trio of beasts retreated. Another time he was with two German tourists when they saw a large male almasty walking into an abandoned house. It turned and scowled at them. The Germans were too afraid to take pictures or follow it into the house. More recently he had seen a family group of them. He had tried speaking to them, but they fled back into the forest.

Rumagha Kulmesov and his wife were a delightful couple who invited us into their house and gave us tea, bread, cheese, and delicious homemade yoghurt. Rumagha had seen a juvenile almasty in his back yard only two years before. One night someone threw a pebble at his window. Thinking it was his son come to visit, he called out, telling him that the door was open.

There was no answer, but some time later, someone knocked at the window. On investigation, he saw what he at first thought was a sack of wool in the corner of the yard. Then he realized it was a young almasty. He didn't get a good look at the face, but he said it was hair-covered, with pale, human-like hands. It made gestures as if it wanted food. Rumagha brought it some bread, which it took. It then made gestures that Rumagha interpreted as meaning that it had a friend who also wanted food. He brought a second piece of bread and left it in the yard. He saw the shadow of the first almasty leaving when he went back inside. In the morning, the second piece of bread was gone.

Rumagha's wife saw an almasty in 1955 at the age of fourteen. She and her family had been deported to Kazakhstan. She had been invited to a relative's house. Upon getting there, she found a number of children huddled in a corner crying. When she enquired what was the matter, one of them told her to peek out of the wooden shutters that covered the glassless window.

In the yard was a weird creature slightly taller than herself. From her vantage point, peeking through a crack, she could not see its legs. The upper part was covered with hair. The hair hung down, obscuring the face, chest, and upper arms. Her description put me in mind of Cousin It from *The Addams Family*. It was slowly moving its arms up and down in the manner of a child imitating a bird. It made a whistling noise like a bird. From time to time, it paused to pick up mud and sling it at the wall and shutters. It was still there when she left some time later. She found it odd that such a "crazy topic" could be of interest to us.

We had heard a recent story concerning a derelict restaurant. A scant few days before, a group of armed police were camping there. When the night air was rent by inhuman screams, they fled. We, armed only with cameras, decided to stay the night there.

The restaurant had been built on the lines of a Balkarian castle. It had a forty-five-foot tower, battlements, circular gardens, and many outbuildings. All were built from great blocks of stone. It fell into disuse in the 1980s. This was a shame, because in its day it must have been spectacular. If someone had the time and money to do it up, it could be a glorious attraction even today.

As it is, it is inhabited only by cows and bats. Most of the rooms were covered in cow dung and we had to search for a relatively clean area to sleep. As the sun set, we set up cameras and a campfire. We put out bait and waited. It was a spooky venue worthy of *Hammer Horror*, *Doctor Who*, or *Scooby-Doo*.

We took turns on watch, waiting for something to come lumbering out of the woods behind the buildings or for a wild scream to pierce the darkness. Nothing came. The camera traps picked up only bats.

In the morning, we met Alexey, Natasha, and Humma as they were going back to the Ukraine. Shortly after we journeyed back to Mineranye Vody. On the way, we passed from the Balkarian area into the Karbodinian. The change was noticeable, with less urban decay and filth. We passed through an area called Bidick, rich in unexplored caves. Anatoly and Grigory were thinking of this area for a future expedition.

And so we returned to England with the samples. The hair from the "nest" Dave found turned out to be nothing more than modern human hair.

So what is the almasty? I believe it exists; both Grigory and Anatoly have seen it. It seems smaller and more man-like than the classic yeti or sasquatch. Grigory thinks it may be a surviving strain of *Homo erectus*. As far as I know, however, no fossil skulls of this species show the distinctive "domed" shape. The almasty could, of course, be a descendant of *Homo erectus* or *Homo habilis*. This species begat many others, such as *Homo heidelbergensis*, Neanderthals, and modern man. Why could it not have another descendant, big, powerful, and adapted for forest and mountain dwelling? The almasty seems very adaptable. Grigory says it can live wherever its ecological twin, the brown bear, can.

On this expedition, I felt closer to my quarry than on any other. It really felt that at any moment I could be staring into the eyes of man's closest relative. Kabardino-Balkaria is a unique place. The almasty population is on the increase, and it seems they are willing to approach human habitation on occasion. There seems a good chance of habituating one and getting conclusive evidence. What this will mean biologically, ethically, and even theologically remains to be seen. Different races of humans cannot seem to get along with each other, much less with a different type of man.

The Mongolian Death Worm

In May 2005, the CFZ undertook our most ambitious expedition to date. A four-man team travelled to the Gobi Desert in search of the infamous Mongolian death worm; a vermiform, desert-dwelling creature said to spit a corrosive yellow venom and held in much fear by the Mongolian nomads, who know it by the name of allghoi-khorkoi (pronounced "olra hoy-hoy"). The team consisted of myself, my two old travelling companions Jon Hare and Dr. Chris Clark, and a new addition to the CFZ expeditionary force: Dave Churchill. Dave was a longtime member of the CFZ and wanted to join the Mongolian expedition when it was first mooted several years ago.

Reports suggested that the death worm emerged after rainfall and lived near sources of water. Therefore, we proposed to try and dam some of the streams in the oasis in order to create localized floods, thus forcing the worms to the surface. Chris also proposed the use of bucket traps. These are buckets buried in the sand. with mesh netting strung between them. The idea is that small creatures would bump into the netting then crawl along it until they reached and fell into one of the buckets. We could then examine them in the morning. He also brought some small mammal traps so that we could try to catch potential death worm prey for examination. We also had some leaflets printed up in Mongolian and distributed throughout the area we were visiting. They explained that a group of British scientists would be travelling through the area in May and offered a fifty-dollar reward for an allghoi-khorkoi specimen.

We flew via Moscow (an airport unique in my experience in not having a bureau-de-change) and on to the capital, Ulaan Baatar. Ulaan Baatar does not look very oriental. It has more in common with Russia and its Eastern Bloc architecture. The buildings are grey or dull brown and functional. The skyline is dominated by a massive power plant that burns coal and pumps it through big, ugly pipe to heat the city. There were a few scattered gurs, the traditional circular, Mongolian tents, in back yards or clustering on the outskirts of the city. It seemed wrong to me that a race of nomads, the people of Genghis Khan, should live in this sedentary manner.

We were met in the airport by Byamba, the director of e-mongol.com, the company with whom we were travelling. Once closed to outsiders during the socialist era, Mongolia is now a popular destination with more adventurous travellers. Byamba told us that a Canadian company owned the mines and power plant and that the boss was called Richard Freeman!

Dave had set up a website, www.cryptoworld.co.uk, in order to chart our adventures. He intended to give regular updates via his laptop, but his USB connection was not working. After checking into the Marco Polo Hotel, we headed to the E-Mongol offices to see if it could be fixed.

We were introduced to our guides Bilgee (pronounced "bilgay"), a stocky, genial chap, and Tulgar (pronounced "tograr"), a slightly shy-looking man in his early twenties. Dave's connector could not be fixed, so he was driven around the city in search of a new one. Whilst he was gone, Byamba tried to contact a man named Boldbaatar who worked as a researcher in the Ministry of Science and Education. Boldbaatar had been researching the death worm for some time, and Byamba wanted to introduce our team to him.

Boldbaatar claimed that he was just leaving his office on a research trip and would be away for over a month. Byamba suspected that he just didn't want to share his information.

That afternoon we visited the natural history museum. The collection of dinosaurs is staggering. They include *Tarbasaurus baatar*, the Asian cousin of *Tyrannosaurus rex*, and *Deinocheirus mirificus*, a huge ornithomimid dinosaur that, until recently, was known only from its vast, eight-foot-long forearms. It is believed to have used its scythe-like claws to collect plant material to feed on. Most impressive was a Velociraptor and a Protoceratops locked forever in a dance of death. The predator is clutching the prey's bony frill, whilst ripping at its stomach with its curved hind claws. The prey is fighting back by biting down on one of its foe's wrists. Both were buried by a sandstorm millions of years ago and preserved mid-fight.

Afterwards we visited a large market area where I was jostled several times. I later found out that some bastard had "liberated" my £300 digital camera.

Dave had no luck in getting his USB to work, so over breakfast the next day, Byamba introduced us Damdin, a friend of his who was a computer whiz.

He worked programming computers for the mining company. He failed to fix the gizmo, but took a great interest in our trip. He told us of a man who had claimed to have seen a creature like a yeti, only smaller, a couple of years ago. However, when the beast was captured, it turned out to have been a monkey that had escaped from the circus!

Another story he had was altogether more interesting. His aunt had told him of a dragon that she had seen in a river in the 1940s. This happened in the north of Mongolia just after WWII. The animal was dead and protruding from a frozen river. At first, I thought it must have been a frozen mammoth, but Damdin said it was long and scaly like a snake. It had been around a hundred feet long, but only the back was visible above the ice. It had been a very hard winter, so the villagers fed on the dead dragon's flesh until the spring thaws washed the carcass away. If only they had kept a few scales!

In Mongolia, dragons are called luu (pronounced "low"). It is believed that they live in heaven and only descend to earth on occasion. They bring rain, and when storm clouds form around a mountain peak, it is called luu hang.

We were introduced to our drivers, Togoo and Davaa, and the vehicles that would be transporting us for the next month. They were tough little Russian all-terrain vans that had been customized into minibuses.

After breakfast, we set off toward the wilderness. On a hill overlooking Ulaan Baatar was a large cairn of stones with a branch protruding. About the branch was tied blue cloth. Bilgee explained that it was an ovoo. Travellers stopped and walked three times round it and placed a new rock upon the pile. This would ensure a safe return. We all added to the ovoo before moving on.

I thought travel in Sumatra took a long time, but it has nothing on Mongolia. The vast distances and total lack of anything approaching roads makes journeys never-ending. We finally reached an area of strange rocks. It looked as if a chocolate ice cream belonging to Godzilla had melted. The brown rocks ran in weird liquid shapes and a sparse wood clustered at their edges.

Bird life was all around. Vast black vultures soared overhead, and wood-peckers flitted among the trees. A pair of noisy ravens eyed our camp with interest. Wherever we went in Mongolia, ravens were present. It became a piece of expedition folklore that they were the same ravens, keeping pace with us like the supernatural ravens of Odin in Norse mythology.

Close to the camp was a very big ovoo with a living tree at its centre. As well as blue cloth, there were prayer flags with the symbol of the air horse (emblem of Mongolia) upon them, or representations of the creatures from the Mongolian zodiac that are much the same as the Chinese.

We ate lunch at a tiny conurbation that had the air of a Mexican border town. There happened to be a Mongolian wrestling bout being held in the village, so we went along. Mongolian wrestling seems to be about getting your opponent off balance whilst grappling with locked arms. You are not allowed to kneel or put your hands on the floor. One poor chap was comically outmatched by the Mongolian equivalent of "Big Daddy."

On the third day, the terrain became more open and the desert gritty. As we drove, Bilgee told me that one of our drivers had been in the area about ten years earlier and had tried to use a well, but found it had been covered and locked. Upon enquiring about it, he was told that a dragon had entered the well.

We located the well. It consisted of a car tire around a hole filled with muddy water. It was only a couple of feet across, so the dragon would have to have been very thin to fit into it.

We enquired at a nearby gur. The lady who owned the gur was most hospitable, and we were soon gathered around drinking salty Mongolian tea. She told us that ten years before, an old wise man had seen a dragon entering the well. He had ordered it to be locked and said that no water should be taken from it. The local children became afraid.

The story got about, and three government officials came to see him; this was in the socialist era and the story was deemed religious, hence against the socialist thinking of the time. The three men poured oil into the well as a punishment to the superstitious people. Soon after, two of the men mysteriously died, and the third remains childless to this day.

The woman did not see the dragon herself, but she told us that it was supposed to change colour like a rainbow. To an evil person, it would appear black.

That night we camped in the shadows of black, twisted mountains that would not have looked out of place in Mordor. We explored them, and I can truly say that the feeling of being watched one gets in eerie places was stronger here than anywhere else I have ever been.

As we drove further south, the land became flatter. I truly doff my cap to our drivers and the amazing way they navigate, without benefit of roads or landmarks. The terrain was monotonous. The area we travelled through was known as the Mirror due to its flatness.

Chris and I had Davaa as our driver. Dave and Jon had Togoo. Togoo drove like a madman and had very little trouble. Davaa drove like an old lady and was constantly running into consternation. Twice we were stopped by punctures, and once we suffered an overheated engine. En route we saw a swarm of black vultures around the carcass of a horse. We also glimpsed some rare black-tailed gazelle (*Gazella subgutturosa*) in the distance.

We stopped to dine at an area known as Long Red Mountain. Once again, the chameleon-like Mongolian countryside had changed. Now it resembled the ruddy surface of Mars. Long Red Mountain is famous for its dinosaur fossils and fossil eggs. It is one of the places palaeontologist Roy Chapman Andrews visited in the 1920s. He was looking for fossil humans, but discovered dozens of dinosaur nest sites. The discoveries continue to this day. Some German tourists found an ancient egg just last year.

We thought we would try our luck and split off to search for fossils. I was looking along the sides of embankments where small landslides had occurred when my eyes fell on a perfect fossil egg. It was a little larger than a hen's egg and more lozenge-shaped. There were tiny cracks running across its dusty grey surface. Excitedly, I reached down to grab it, with thoughts of presenting it to the natural history museum in Ulaan Baatar. It instantly crumbled to dust in my palm. It was nothing but desiccated camel shit!

I was laughing at this when the sandstorm blew up. It came from no-where, a screaming red storm that lashed the sand into a frenzy and tore at the skin and eyes. I stumbled back to camp and we drove away. It was like driving through aggressive fog.

We finally drew clear of the storm and passed a well. Around it were gathered nomads with herds of sheep, goats, camels, and horses. Water was drawn up via a bucket on a rope attached to a wooden, hinged, weighted tilt. The weight counterbalances the water bucket, making drawing it up far easier. It is a tradition in Mongolia that, when travellers pass a well in use, they stop and help draw water for the livestock, so we all took turns filling the troughs for the animals.

One of the nomads offered me a ride on a camel, a grumpy-looking beast that was kneeling in the dust. I accepted and had begun to mount it when I committed a dreadful faux pas in camel etiquette. When mounting the ship of the desert, one is meant to throw one's leg between the beast's humps, then insert one's other foot into the stirrup on the near side. I put my foot in the stirrup before throwing my leg over. The indignant camel screamed in fury and turned its head round to face me with a look of pure hatred. Bearing nasty yellow teeth, it reared up and threw me to the ground, then attempted to run off into the wilderness. It took three men to regain control of the beast. It must have got the hump.

We decided not to camp that night and stayed with Davaa's family in Mongolia's second city, Dalanzagad. The Mongolian equivalent of Birmingham made Ulaan Baatar seem like Venice or Prague, but it did have an internet café where Dave could update the website.

Jon and Dave were accosted by a drunk who told them that they were fools to hunt for the allghoi-khorkoi as it had "killed thousands of people."

On the outskirts of Dalanzagad lived one of the witnesses who had seen our posters and contacted Byamba. Luvsandorj was a ninety-year-old former policeman. He lived in the Gur district of the city. He invited us in and, following tradition, offered us some snuff from a bottle. I had never had snuff before, and thinking it churlish to refuse, I took a hearty pinch and snorted it up one nostril. I almost fell over with a fit of sneezing, much to the amusement of our host.

It was 1930 when Luvsandorj saw the death worm. He was fifteen back then, and had been tending to cows when he came across a two-foot-long, reddish-brown creature in the desert. It was about four inches thick, and he could see no eyes or mouth on it. It was sausage-shaped and moved slightly from side to side. He ran to tell his parents, who warned him not to go near it, as it was deadly. He drew a picture of what he had seen, a crude sausage shape. He provided us with the names and addresses of other people he knew that had seen the creature.

As we drove into the desert, we saw what looked like a huge, shimmering lake overlooked by mountains. As we got closer, we saw it was a titanic mirage. We camped in the freezing desert beneath weird black rock vistas.

Next day we travelled deeper into the wilderness and tracked down one of the people on Luvsandorj's list. Juuraidor was a seventy-year-old camel herder who saw the worm in the 1950s whilst searching for lost camels. His description tallied with that of Luvsandorj's: brown, two feet long, and with snake-like scales. He had heard that the worm was dangerous, so he ran away. The encounter was in June. He also told us of a man who had put a death worm on an iron plate. The plate had turned green. Another man that he knew had wrapped a dead death worm in three layers of felt. The worm shrivelled up like a piece of leather, and the felt turned green. Both incidents had been long ago, and no remains had been saved. Later that day, we saw a group of rare Mongolian wild ass (*Equus hemionus hemionus*) galloping across the dusty horizon.

We had now entered the Gobi proper. We made camp in a rocky valley. Two men drove into the camp the following morning and introduced themselves. One was a grizzled park ranger, the other a younger man with prominent golden teeth named Nyama. The latter had seen the death worm on no less than three occasions.

The first was in 1965, when he saw the creature's head (presumably) protruding from a hole in the sand. The following year, he saw a specimen in the process of swallowing a mouse. Finally, he actually killed a worm in 1972 by throwing a rock at it. Some Russian scientists who had been in the area studying snakes took the body away. It probably resides to this day forgotten in some Russian museum basement.

Nyama said the worm eating the mouse was grey and ten inches long. The other two were brown. The one he killed was between eighteen inches and two feet long. They moved with a caterpillar-like motion. The sightings occurred in a place called Dun-dus. He also heard tell of a death worm killing a child by spitting venom, but could not confirm this.

The park ranger and his family lived in a nearby gur. His wife had seen the worm just three years before in an area close to the Chinese border. He invited us into his gur whilst we waited for his wife. He had an impressively large collection of goats and some huge, savage-looking guard dogs. Despite an intimidating display of barking, the dogs were actually quite friendly. I found this to be the case right across Mongolia.

He offered us a drink of fermented camel's milk whilst we waited. It tasted like fizzy, alcoholic yoghurt. It made a nice change from salty tea, and I drank two bowls of it, an action that I would later come to regret.

When Sukhee, his wife, turned up, she agreed to take us to the spot where she had seen the creature and remained with us for several days' hunting.

As we were close to the Chinese border, we were stopped several times by amiable but Kalashnikov-toting guards who checked our passports. One had heard about our expedition on the wireless. Byamba pointed to a nearby range of hills and told us "That marks the border with China." The hills were just twenty miles away.

The ground was now sandy and full of rodent burrows. A low forest of saxaul grew. As we wandered through the stunted woods, Sukhee saw something and began to dig. She pulled a light yellow, ugly root from the ground. It was waxy in appearance and was covered with soft spikes. It was the root of the goyo plant and much admired by nomads. We shared the root, which tasted surprisingly good, rather like a cross between banana and celery. Where the idea that goyo is poisonous originated, I have no idea. It made very good eating.

She led us to the spot where she had seen the strange beast. She had been herding cows with her son when she saw an eighteen-inch, grey, worm-like thing slither out from a hole. Her son threw a rock at it, and it slid into some bushes. She ran away. It had been in September, and it was very hot, about forty degrees Celsius.

Sukhee told us there were two worms in the desert. One was the all-ghoi-khorkoi or intestine worm. The other was called the temrenii suhl or camel's tail. This was smaller than the death worm and grey rather than red-brown.

After a search of the area, we decided to make camp. Just after lunch a particularly savage sandstorm blew up. It tore down our tents and blew over our tables. We retreated to the safety of the minibuses, but my stomach was bubbling loudly on account of the fermented camel's milk. I had to brave the stinging sands and make a run for a bush. Having explosive diarrhoea in the middle of a sandstorm is not something I would like to repeat!

Sukhee told us that the storm would last for the whole night. It was impossible to camp, so she invited us back to her gur. The family had two gurs and allowed us to sleep in one of them.

Dave's laptop caused much interest as he had moving film with sound on it. The whole extended family crowded into the gur to see it. It was like a small, crowded, circular cinema. Many of the Mongolians (including Bilgee) had never seen the sea and were fascinated by Dave's film.

We travelled back to Dalanzagad and as Davaa's family was away, we checked into an ugly hotel and spent the night at Dalanzagad's answer to Stringfellows.

Next day, we drove deep into the Gobi and headed for the frozen gorge, a river that never thaws. We visited the Gobi Museum, surely the remotest museum on earth (unless there's one in Antarctica). Among the wide array of stuffed animals was a carving of the death worm. Adam Davies had seen and photographed this a couple of years earlier. It resembled the witness descriptions apart from having clearly visible eyes.

The creature was labelled a tartar sand boa. Via Bilgee, I enquired as to why this was. The museum guide said that the carving had once been labelled as an allghoi-khorkoi, but the label got tatty. He was at a loss to know why it had been relabelled a sand boa.

The tartar sand boa (*Eryx tatricus*) does bear a slight resemblance to the death worm descriptions. It is a brownish or greyish, chunky snake that can reach four feet in length. The end of its tail is noticeably thick. However, it has a clearly defined head and visible eyes. It ranges from the Caspian Sea to Western China. It probably occurs in Mongolia as well. The sand boa is a constrictor and does not produce venom, let alone spit it.

We camped in a valley full of pikas (*Ochotona pusilla*), small lagomorphs related to rabbits and hares. The next day we set off for the gorge. Yolyn Am is a frozen river, protected from the sun's rays on each side by sheer five-hundred-foot-tall cliffs. At one time the river never thawed, even at the height of summer. But from 1980 onwards it has begun to thaw in summer. Perhaps this is a further indicator of global warming.

When we were at Yolyn Am, the river was frozen. The ice was thick enough for us to drive the minibuses along it, until a sheet gave way with a sickening crash and engulfed the back wheel of the one I was riding in. We had to tow it out.

Continuing on foot, we found an ovoo built on the ice and some amazing ice caves. Where the lower layers of ice had melted, they left blue and silver caverns in the frozen river. We crawled inside for a closer look. Weird fissures and troughs had been carved out of the ice by the wind.

In a small cave by the riverside, Chris left a geocache, a plastic container with the details of our location via satellite and details of the CFZ website so that whoever found it could get in touch.

Bilgee said that this area, an eastern outcropping of the Altai Mountains, was one of the best places in the world to see snow leopards. On three of his four trips there, he had seen one, and Togoo has also seen one killing a wild sheep. Sadly, we were not so lucky.

We met a group of men selling stone carvings. One of them, an incredibly talented man called Baiar, carved me a dragon in one hour flat.

A massive sandstorm blew up that evening, turning the sky a reddish grey. Deciding not to camp, we stopped in a small town and slept in a minute hotel with unreliable electricity and a constantly banging door.

Striking on in the morning, we reached Noyon Sum. A sum is the Mongolian equivalent to a British county. Some sums are the size of Scotland. In the Sum Centre (the largest and most important town in the sum), we had a meeting with the governor of Noyon. He had never heard of the death worm but received the leaflets Bilgee had distributed. The governor did some digging and found out that in 1955, a man herding sheep had come across a two-foot-long, grey creature with no discernible head or tail. The man fled in terror. He now lived in Dalanzagad.

The governor had found more stories. His driver told us that in the 1960s his mother-in-law had seen a death worm. It was light grey and making holes in the sand. She ran away from it. The governor had found another local man named Damdin who had seen the creature in 1954–55. He gave us his address and we thanked him heartily and went on our way.

After a roundabout drive, we found the gur belonging to Damdin's sister. She told us that her mother had seen it as well, also in 1955. She had been present at the sighting but was too young to recall it. Her mother said it was two feet long, brown, scaly, and as thick as a gur support pole (about five inches). She gave us directions to her brother's gur.

We found Damdin's domicile and he welcomed us in. He told us he had been out tending camels in May 1955. At about ten in the morning, he saw a death worm. It was brown, two feet long, and about two inches thick. It made no movement. He ran to tell his parents and they warned him it was venomous. He returned to the spot he had seen the creature in, and it was gone.

Damdin's family had been so frightened by what their son had seen that they packed up their gur and moved. He said that lots of families moved after seeing a death worm. He heard tell of it killing animals by spitting at them.

He took us to the area where he had seen the worm. The marks of his family's old gur were still visible in the gravel. The area was much disturbed by camel tracks. We felt it unlikely that the worm still inhabited this part of the desert. Another storm forced us to spend the night in the gur of one of Damdin's friends.

We moved on again to stay in another gur, deeper in the Gobi. The one we were supposed to be staying in was flooded, but a second one was available. There was rancid surface water all over the area, and sulphurous-smelling salt-caked grasses.

The valley was overlooked by an old telecommunications tower from the socialist era. We explored some crumbling barracks dating from the same period. Behind these was an old Buddhist temple with ornately carved pillars, some of the original colours still visible. The temple had been destroyed in the anti-religious purge of the 1930s. What was left had been incorporated into the barracks. Deep in the bowels of the temple, beneath some old floorboards, I found an offering, mouldering money wrapped in a blue cloth. It seemed that even at the height of socialism the faith had not been totally extinguished. I added a little of my own money and reburied the offering.

Further beyond the temple was a weird landscape, an oasis of mossy, green, humped tussocks that looked more Irish than Mongolian.

The next day Bilgee took our passports to be checked at a military base. We were once again drawing close to the Chinese border. He returned with a retired Mongolian Army colonel called Hurvoo, who wore a broad ten-gallon hat. He had once been in charge of a base called Ovootin Otriyad.

In 1973, he had been patrolling an area called Ulann Ovoo on motorbike. It had been in May and at sunrise. He saw what looked like an old tire in the desert. It was some sort of animal lying coiled up. His description was by now familiar: brown, two feet long, scaly, and sausage-shaped. He did say he saw light playing across it, like electricity or light reflected from a mirror. This may well have been the rising sun reflecting off its scales. It had been raining and the worm was wet.

Hurvoo watched it for half an hour, and it did not move. He drove off to get a camera, but on his return it had gone. One year later a solider reported that he had seen an identical animal. Hurvoo investigated but found nothing. He believed the worm came out after rainfall.

We drove out to the area after checking in at the military camp again. We set bucket traps in the scrub. These consisted of a series of sunken buckets connected by netting supported by struts. The idea is that a creature bumps into the netting and cannot continue forward. Ergo it runs along the netting until it comes to a bucket and falls in. Then you have a specimen. We also set baited small mammal traps to see what possible death worm prey might be in the area.

That night we stopped at the military base. A thunderstorm occurred during the night with heavy rain. The next day we rose early to check the traps. Both the bucket traps and the small mammal traps were empty.

We moved on to Gurantes Sum and stopped in a hotel shaped like a huge concrete gur. We had a meeting with the governor of Gurantes, Deevat Serem, who had distributed our leaflets in the area. He had a witness named Khuuhengaa with him. She had seen the worm in the 1980s when she was a girl. She could not recall the exact year, but it had been in summer. She was staying with her grandfather, who called her to see it. The worm was fifteen inches long, brown, and with no discernible head or tail. Her grandfather told her it was venomous, and she was afraid.

Deevat told us of another sighting that had occurred just in the previous year. A man had been cutting reeds at an oasis called Zulganai. A man cutting grass lifted up the worm on the end of a stick and threw it away. Another man had seen the worm at the same oasis and claimed he could identify its tracks and burrows. Deevat was sure the worm exists and told us that in these days it is seen less often. This is not because it is getting less common, but people are now travelling by motorbike rather than by horse or camel. Also, people are moving to towns, cities, and areas of sedentary residence, rather than moving about as they used to. Hence the death worm is encountered less often. We identified three oases that we thought might be promising places to look, given that the death worm was found close to water. He thought that the dragons seen in wells were just metaphors for poisoned water or ways of keeping people from drinking bad water.

The morning found us back on the trail again. We tried to locate the man who could identify death worm tracks, but he was not at home. We found the gur of Batdelger, the man who had seen the worm in 2004 at Zulganai. His wife and son also saw the creature. They had been cutting grass to feed livestock at the time. His description differed slightly from the others. The worm he saw was fifteen inches long and brown. It had a squarish head and what looked like large eyes, but these may have been part of a pattern on the skin.

He did not think it was a snake, as it was too thick. His son lifted the worm on a branch and cast it away. It felt very heavy.

We spent the night in a gur belonging to the family of the former local governor. The next day we spoke with his wife, who had seen an allghoi-khorkoi in 1957 in an area close to the Chinese border, near where Colonel Hurvoo had seen it. Once again, our suspect was fifteen inches long, brown, and had no clear head or tail.

The ex-governor himself spoke to us and said he knew a man who had seen three large snakes some years ago. The biggest was six feet long and had a head shaped like that of a sheep. All three sported horns.

There are horned snakes known to science. They include the rhinoceros viper (*Bitis nasicornis*) and the horned viper (*Cerastes cerastes*). Their horns are in fact modified scales. However, none is known from Mongolia, and none reach six feet in length. We were unable to locate the witness.

A long, boring drive followed. At last we came to an area of spectacular cliffs and mesas. They looked like gigantic, petrified flames. Between these spectacular edifices was Wall Canyon or Zuun-Mad, the first of several oases we intended to visit.

Water had cut low but sheer cliffs in the sandstone. A stream ran through them, surrounded by poplar trees, saxaul, tamarisk, and other plants. We split up to explore and I followed the path the water had cut into a narrow, steep side valley. A short-eared owl (*Asio flammeus*) exploded out of a bush about five feet away, causing me to jump out of my skin.

Later we set up the bucket and small mammal traps again. The physical nature of the oasis with its steep sides made the damming plan impractical. In the twilight the colour of the rocks changed hue as you looked on. It was a spectacular display that looked more like special effects than a natural occurrence.

We decided to explore Wall Canyon after dark and set off with torches. Dave's beam picked out a pair of large, luminous, green eyes. We tried to get a closer look, but their owner had vanished.

At daybreak we examined the traps. Apart from hordes of ants they were empty. We all saw the owl again. Short-eared owls, unlike most of their kin, will often hunt in broad daylight. The riddle of the green eyes was solved that night. They belonged to a Przewalski's skink gecko (*Teratoscincus scincus*). We saw dozens of these attractive little lizards, their outsized eyes glowing green in the night. We caught several. Their luminous eyes give the illusion of a much larger creature.

Next day we moved on. Another epic drive brought us to the largest collection of vegetation we saw in the whole Gobi, the oasis of Zulganai. This was a mile-long march with steep cliffs at one side and a kinder slope at the opposite side. It was filled with reed beds as tall as a man. At one end a narrow strip of woodland erupted, and the oasis formed a second smaller pool. The water course ran on to be lost in the desert. Large, savagely horned bulls pranced through the reeds adding a little extra spice to the exploration.

We descended into the reed beds. There was much bird life, including demoiselle cranes (*Grus virgo*), spoonbills (*Platalea leucorodia*), and white storks (*Ciconia ciconia*). We saw the strange flowers of the goyo erupting from the earth. These consist of reddish, phallic-shaped buds that bring forth masses of tiny violet and blue flowers. They look a little like lupins.

We made camp and, as dinner was being prepared, we watched a whirlwind forming in the distance. It had started as nothing more than a dust devil but rapidly grew in size. We watched it with interest, filming it as it slowly drew closer and built in size and power.

It approached the valley that curled up over the lip of the cliff and was upon us in seconds flat. Screaming like an angry djinn, it tore through the camp, shredding the tents and hurling dinner to the four winds. I was slammed against one of the minibuses and found myself in the eye of the twister. Togoo was entangled in the remains of a tent and dragged across the desert.

Then, just as swiftly as it had come upon us, it was gone, sweeping out over the dusty plains and toward the horizon.

The camp had been totally destroyed and night was falling. We had no choice but to drive seventy-five kilometres back to Gurantes and spend the night in the stone gur. It was as if the desert was rallying its powers against us, jealous of its secrets.

We drove out to the third oasis, Hermen Tsav, the next morning. This was a small wood of poplar trees. It was much drier than the others and far more used by humans. An animal corral had been erected at the centre of it. There was little wildlife. Thankfully we had backup tents. Jon and I started one with pictures of grinning clowns on it. Adam Davis had told us of this very tent the year before.

That night Tulgar soundly thrashed Jon and me at Mongolian wrestling and Bilgee told us a strange story about a haunted gur. Back in the 1970s, a sombre socialist newspaper reported that a Russian geologist had been attacked by a ghost in an empty gur. Gurs that are not in use are often stocked up with fuel, food, and water for fellow travellers to take advantage of. In a land as inhospitable as Mongolia, this is a necessity rather than a kindness.

The geologist and a colleague were staying in such a place one night. They had been playing cards. The geologist looked up to see the inside latch on the door turn and open by itself. His friend did not seem to notice it. Presently a woman in a red deel (a long heavy Mongolian coat) entered and tried to strangle him. His companion was oblivious. He finally managed to kick over the card table, whereupon the ghost vanished and his friend looked up.

In the morning we drove back to the second oasis, from which the whirlwind had driven us. Exploring the second smaller pool, we discovered the tracks of several wolves going to and from the water source. We followed the tracks and the stream that fed the oasis into the desert. We tried to find its source, but it became lost among the trackless sands, and we were pressed for time.

We found another smaller pool that had methane bubbling up from it and releasing foul smells. That afternoon Bilgee discovered that one of the vans had a badly damaged axle and it had to be driven to the nearest town (Sevree Sum) for repairs.

Next day we took the long dull drive to Sevree Sum. We saw a tornado on the way that dwarfed the one that had destroyed our camp. We filmed it from a distance this time.

When we arrived in town, we checked into a tiny hotel. The matronly owner cooked us some delicious meat dumplings. The governor had heard of our arrival and invited us into his gur. His name was Tserendorj and his hospitality overwhelmed us. He told us that he was delighted that we were visiting him and that he did not think that scientists from abroad would be at all interested in meeting him. He seemed to think we were far more important than we actually were.

Tserendorj was a mine of information and, together with his friends and family, we consumed much vodka during a fascinating evening. He related a story of an old man who had seen the worm in 1957 in an area of the Gobi now owned by China. It looked like a length of blood-filled intestine. The same old man heard of a fellow who prodded it with a horse goad and the end of the goad turned green. Both horse and rider died. This is reminiscent of stories of the basilisk in medieval Europe. Another man had seen it during the 1950s. It slithered out from under a rock and the witness threw a stone at it. The worm retreated back under the rock.

The governor introduced us to a ninety-three-year-old man whose grandfather had seen the worm in the nineteenth century at the oasis we had just left. He thought the use of motorized vehicles and the population moving to towns was the reason that the death worm was now seen less often.

He did not buy the idea that well-dwelling dragons were metaphors for bad water at all. He insisted that they were real creatures (and I am inclined to agree with him). He told us of a doctor from Ulaan Baatar who had seen such a creature in a well just last year. This had occurred in Bulgan Sum. The doctor described what looked like a green-scaled Chinese dragon coiled at the bottom of a well. Understandably he was shocked. This was no nomad or peasant, but an educated man.

The governor had also spoken to a man who had seen a horned snake in his youth. It was over six feet, five inches long and sported two horns. He found it outside of his gur as a child. It did not seem aggressive, and he played with it.

Years later, the man's wife died after a five-year illness. A shaman told him that it was because he had touched God's sacred creature as a boy.

The governor invited us back next year to join in the celebrations for the eightieth anniversary of his Sum.

On the morrow, we were back on the road. We stopped at Bulgan Sum but were, frustratingly, unable to locate the doctor. A beautiful teenaged girl was selling unusual stones she had found in the desert. I brought a couple of pieces of petrified wood from her.

That evening we visited the famous flaming cliffs where Professor Roy Chapman Andrews discovered the first dinosaur nests and eggs in the 1920s. We stopped at a nearby holiday camp consisting of rows of gurs.

The expedition was all over bar for the shouting now. The next couple of days was nothing but a dull journey back toward the capital. We had a brief look at a Buddhist monastery that had escaped the anti-religious purge of the socialist days by becoming a warehouse. Once Mongolia had thrown off the yoke of communism, the temple reopened. There were interesting carvings of daemons. Before the coming of Buddhism, the Mongols were animist.

When Buddhism arrived, it absorbed the old animist gods. It turned them into daemons, but ones that had converted to Buddhism and were ferocious, zealous guardians of the faith.

We contacted Byamba and asked him if he could try to contact the doctor who had seen the dragon or find out his address in Ulaan Baatar.

We had one day staying in the "melted chocolate" valley. Dave almost got himself killed climbing up a mountain.

Finally, we made it back to Ulaan Baatar and checked into the Marco Polo Hotel again. The next couple of days were spent around shops, museums, and temples. We visited the library to see if we could turn up any information on the death worm. They had nothing on the worm but did have a massive block of stone intricately carved into entwined dragons. It formally sat upon another block onto which had been carved with an untranslated script. It had been found in the desert some years before.

We were unlucky insofar as nature was against us. May 2005 was colder and windier that the norm. We intend to return one day in more clement weather.

Byamba is still trying to track down the doctor who saw the dragon. So far, he has had no luck.

I believe that the death worm is one of two things. It could be a worm lizard or amphisbaena, a group of primitive burrowing reptiles that are not worms, snakes, or lizards, but are related to the latter two. Poorly studied, little is known about these creatures, but they resemble the descriptions of the death worm. Indeed, they take their name from a legendary snake, the amphisbaena, that had a head at each end of its body.

Another possibility is that it is an undiscovered species of sand boa, a sausage-shaped constricting snake often found in arid climates.

Neither worm lizards nor sand boas are venomous, but strange beliefs can grow up around harmless creatures. In the Sudan, the natives believe that the sand boa is so deadly that one only has to touch it and the venom will soak through your skin and kill you. They call the sand boa the apris and go in great fear of it, despite that fact that in reality it is harmless.

There could well be a species of horned snake in Mongolia, probably a form of large viper. As for the dragons, the massive international accumulation of sightings and folklore convinced me of their existence a long time ago.

I have emailed Byamba some pictures of worm lizards and sand boas to show to witnesses he speaks with. Perhaps they will see something they recognize. For the time being, the wind-haunted, desolate Gobi is keeping its secrets.

The Giant Anaconda

The Centre for Fortean Zoology's expedition to Guyana had its genesis with an entry in Michael Newton's excellent *Encyclopaedia of Cryptozoology* on crypto-tourism. He mentioned a company called "Guided Cultural Tours" that offered expeditions in search of giant anaconda in Guyana.

I contacted Damon Corrie, who ran Guided Cultural Tours. The company specialized in showing people the true Guyana of the native peoples. Damon himself is a chief of the Eagle Clan Arawak Amerindians. As well as being a well-respected figure to the native peoples, he also is a conservationist who breeds and studies Guyanese reptiles and invertebrates.

Damon told me that only the year before, at a remote pool known as Corona Falls, a gigantic anaconda had been seen. He had spoken with the hunters who had seen the beast. They told him that it was so large they had fled from it. When he asked how big the snake was, one of the men pointed to a thirty-foot palm tree. He told Damon that a dead tree of the same size had been lying in the water. The anaconda was crawling over it and its head and tail extended beyond the ends of the tree. This would make the snake around forty feet long.

Damon also mentioned the di-di, a large hairy creature seemingly akin to the yeti or sasquatch, seen in Guyana, as well as more vague stories of dragons said to inhabit the mountains. The CFZ decided to mount an expedition in November of 2007 in search of these creatures.

For the first time ever, we were able to secure some funding. Sam Brace, who worked for the computer games company Capcom, tied in our expedition to the release of Capcom's game *Monster Hunter Freedom*. Capcom kindly donated £8,000 toward the project.

As well as myself, the team included Dr. Chris Clark and Jon Hare, who had been on several past expeditions, and Paul Rose, a.k.a. Mr. Biffo. It was Paul's first such excursion. He is a TV writer, author, and most importantly the man behind the sadly defunct "Digitizer" on Teletext. It was collection of bizarre non sequiturs that had Jon Downes and me in tears of mirth for several years.

We were met at Georgetown airport by Damon, who took us to his native village of Pakuri. It was a long and bumpy ride on an open-backed truck to the village. Pakuri is also known as St. Cuthbert's Mission, but Damon, who is a champion of Amerindian rights, encourages all to call the village by its original name. It takes its name from the pakuri tree. Damon showed us the last such tree in the area, the others having long since been cut down. I made a point of always calling the village Pakuri, as I hate missionaries destroying native culture with their pernicious twaddle.

Pakuri seemed like a content and stable community, unlike the filthy and crime-ridden Georgetown. We took a swim in a nearby creek. The water was stained red by the tannins from the leaves of the plants along its banks. Once in the wine-coloured waters, it gave the illusion of turning your skin red. Damon told us that small caiman and anaconda were sometimes seen in the creek, but thankfully, the infamous genital-invading candiru (*Vandellia beccarii*) was absent from the waters.

Whilst in Pakuri we were told of a di-di encounter that occurred only two years before. It happened in another Amerindian village some thirty miles north of Pakuri. Two children, a boy and a girl of about twelve, were walking home from school across the savannah. What the boy described as a "huge hairy man" stepped out of a stand of trees and grabbed the girl. She was never seen again. There was no police investigation. This is unsurprising, as the government of Guyana seems to care very little for its native peoples.

A man from Pakuri had seen a di-di from the back as it walked away from him several years previously. Unfortunately, he was not in the village at the time, so we could not interview him.

Damon showed us a rainbow boa (*Epicrates cenchria*) he had captured. It was a rusty pink colour phase that I had not seen before. We were also shown a new species of green scorpion discovered by Damon. It was so new that it had not been officially described or given a scientific name yet.

Damon's brother-in-law Foster told us that, several years ago, in a water-course a few miles from the village, he saw the trail of a big anaconda. Judging by its width, its maker would have been far larger than the twenty-foot stuffed specimen in the National Museum in Georgetown.

We travelled back to the unpleasant environs of Georgetown to catch a bus inland to Lethem, where the expedition proper would begin. When the coach turned up at the shabby little station, our hearts fell. It looked as if it were held together by rust. I seriously doubted that the malodorous vehicle could make the twelve-hour journey. The seats were appallingly uncomfortable, and the only air conditioning was open windows. It turned out I was right about the bus, and it broke down. We waited four and a half hours for a replacement to arrive. Thankfully it was a little more comfortable, but that didn't stop this bus breaking down for a while as well.

We drove through farmlands, deep jungle, and finally onto the grass-lands. We caught a ferry across the Essequeibo River, the largest in Guyana. Damon told us that there were islands in the wider parts that were larger than Barbados!

We met a pleasant American girl called Rhiannon on the ferry, who was working for the VSO. She had been in the country for many months.

We arrived in Lethem and took an open-backed truck out onto the savannah. On the way we saw many birds, such as caracara, egrets, and jabiru. The landscape was very sparse in trees, but scattered with termite mounds that looked for all the world like Christmas trees constructed out of mud. We reached the tiny village of Toka, where we picked up some more guides and porters as well as several teenaged girls who were to cook and wash for the expedition. We then started out toward the village of Taushida. Unfortunately, we were hiking at noon when the sun was at its most ferocious.

The heat on the grasslands of Guyana was quite unlike anything I had ever encountered before. In comparison, the heat of West Africa seemed like a chilly winter's day. In Indonesia and Thailand there was shade, but that is a commodity lacking in this part of the world. The relentless heat and lack of shade affected me badly, and I suffered from sunstroke. Several times I collapsed on the way to Taushida. The six miles seemed more like sixty. I had to take many rests, but during one of these, I saw a hummingbird at close range. One guide, Joseph, told me that a di-di had been seen in the area. It resembled a huge white man covered with hair. It had been seen peering through some vegetation in the mountains.

When I finally arrived, I was able to wash in a stream close to the little village. The cool water was a blessed relief. We relaxed as tiny cichlid fish (probably bucktoothed tetra, *Exodon paradoxus*) nibbled on our toes. That night, as we made camp in the village, a bush fire sprang up on the far side of the creek. It resembled a serpent of fire as it grew like some medieval salamander uncoiling in the night. Thankfully, the flames did not reach over the water to menace our camp.

Damon told us that, a couple of weeks before, he had seen some strange lights in the sky above the mountains. A light that resembled a bright star had appeared and had seemingly broken into several smaller lights. They had remained visible for some time. Could the place be a window area? As I pondered the question, a meteor flashed across the sky, burning up in the atmosphere and undimmed by light pollution.

Unfortunately, we were told that Corona Falls was a full seventy miles away as the crow flies! There was no way we could walk overland in the heat. It would mean walking twenty miles per day there and back. Six miles had almost killed us. We considered renting a helicopter when we got back to Lethem. But in the meantime, there was much to see in the immediate area.

In the morning we rose early to climb up Makuzi Mountain. Five years earlier, a hunter named Moses Iza had stumbled across an amazing discovery in a tiny cave atop the mountain. In the cooler morning climate, the climb was relatively easy.

The remains sat in a shallow cave. They were in a large earthenware pot. The scattering of flat rocks suggested that the entrance might once have been covered. There was a whole skull of a boy of between nine and twelve, as well as the jaws and ribs of an adult man. The ribs and jawbones were in a smaller container within the large pot. There were also small beads and the tooth of a peccary. The tooth had a hole in it, suggesting that it may have once been a necklace.

Damon did not know exactly how old the remains were. They could have been pre-Columbian, over five hundred years old, or as recent as the end of the Amerindian wars around a hundred years ago.

We all took a drag on a cigarette (despite the fact that everyone except Chris was a nonsmoker), as this was a custom to show respect to the dead.

The older man was obviously someone important in order for his remains to have been buried in such a prominent place. Perhaps he was a shaman or a chief. The boy may have been a sacrifice. The long and short of it is that we don't know.

Moses related that a di-di had been seen walking across the mountain about ten years before.

Another guide, a local hunter named Kennard Davis, told us a story that his father had related to him. It happened in the 1950s. A man had been hunting and was coming home over the mountains. The mountain pass was quicker than walking all the way around. He was holding two wildfowl he had killed, one in each had. As he neared the top of the mountain, he looked up and saw a huge hairy man asleep in the trees. He seemed to be using the vines like a hammock.

The man was so frightened that he ran all the way to the bottom of the mountains, still clutching the birds. When he returned to his village, he fell ill and believed that the di-di had put a spell on him. He consulted a shaman who went into trance to contact the di-di. The shaman told him that the man had frightened himself into sickness. The di-di lived on the mountain and had a wife and daughter who lived on a neighbouring mountain. They did not harm people.

Kennard had never seen a di-di himself, but he did tell us of one strange creature he encountered. Up until the 1970s, a tiny, red-faced pygmy was well-known in the area. He was hairless, naked, brown-skinned and about 3 to 3.5 feet tall. His red-painted face always wore a strange grin. He would leap out of the bushes, grinning at passers-by and scaring them, though he never did anyone any harm. Kennard's uncle had a motorbike and the little red-faced man would often hop onto the back and catch a ride. He always leapt off at the same spot, which Kennard's uncle assumed was his home. People left gifts of tobacco out for him, according to Damon.

The food on the expedition was generally good. We dined on chicken, rice, fish, and cassava. The latter (*Manihot esculenta*) is a major source of carbohydrates. The native peoples shred it, then smoke it, to remove the toxins. It is squeezed through a wickerwork tube, then dried and pounded into a granulated form. Cassava is remarkably filling. A small portion can keep you going for a whole day. It can be eaten in a soft form that is akin to kous-kous or a hard, granular form that is not unlike granola. In both of these forms, it is quite palatable. However, when turned into cassava bread, it has the taste and texture of chipboard.

Later that evening, as the sun became less fierce, we traveled in the opposite direction to visit Tebang's Rock. This was a thirty-foot-tall pillar of rock that stood on the savannah. Kennard told us of Tebang. He was a little man who walked around at night touching children in order to transmit disease. Once a child had succumbed, Tebang would fashion a flute from their bones and play it atop his rock. He was still supposed to be seen on moonlit nights, whistling and shrieking. He seemed to be totally different from the red-faced pygmy Kennard had mentioned previously. Tebang recalls the African goblin Tokoloshi, a horrid creature with an outsized head that wandered the night transmitting illness to children by touching them.

There was another mysterious mountain in the area. Damon was told by a man who was climbing it that he had almost been sucked into a cave near the summit by a dragon. He did not see the dragon, but a great sucking force came from the cave's mouth and almost pulled him in. Damon thought that the force might have been wind blowing through the mountain, if it was hollow or had a network of caves. The cave might have acted like a wind tunnel under the right conditions. Where the idea of a dragon came from is unknown; there are Chinese immigrants in Guyana, but they are few in number. Could their culture have reached these remote areas? As I have noted elsewhere, the dragon is a universal monster and cannot by written off lightly as a mere myth. As the dragon mountain was ten times as high as the mountain where we had seen the burial, we decided not to climb it in the current climate!

The next day we returned to Toka. The walk went well until the sun reached its zenith, then sunstroke began to kick in again. Once we reached Toka, we all slept in the shade for several hours. On all of our other expeditions, we tended to rise at about seven o'clock and then trek through the day till about seven in the evening. In Guyana we could not do this on account of the heat. We were forced to stay as still as we could for the middle portion of the day due to the heat. This meant that much valuable time was lost.

We caught a small open-backed truck and travelled a few miles down a dusty road, then walked to our next destination, Crane Pond. Here there were anacondas. Not monsters like the one at Corona Falls, but average-sized specimens. Still, it would be interesting to see one. It was cooler now, and as we walked across the savannah we came across a female giant anteater (*Myrmecophaga tridactyla*) with a baby upon her back. The spectacular beast began to lollop away from us, until Damon chased it and drove it back toward us, allowing us to film it. It was a dangerous endeavour, as the anteater is armed with formidable claws that it uses not only to rip open termite mounds and anthills, but also in defence against predators. Anteaters have been known to disembowel humans who attack them. We got some wonderful film of the anteater before she galloped over the horizon.

We set up our tents as the sun began to fall. Kennard told me that Crane Pond was once thought of as a dragon's lair. There was an old saying, "Do not sleep too deeply at Crane Pond or the dragon will take you." He said that back in the 1950s, when cattle ranching was still big business in Guyana, a group of cowboys had camped at Crane Pond for the night. During the night they heard a huge animal rising from the water and could hear it breathing. The cowboys panicked. Some gathered up their horses, whilst others fruitlessly fired their rifles at the noise. They beat a hasty retreat.

Could the dragon have been a giant anaconda? Anacondas do make a strange sound when breathing. It has been likened to snoring.

That night I was badly bitten by insects. In the cool morning before sunrise, we woke and at first light set out in search of anaconda. We found large furrows filled with water going to and from the swamp. They were the trails of anaconda. By the look of them, the snakes would have been fifteen to seventeen feet long, average size. Kennard saw a baby anaconda of around four feet slip into the water, but the larger ones eluded us.

We followed a line of trees along a partially dried-up creek but found no anaconda. As the sun was getting higher, we decided to return to camp. Kennard traveled further down the creek, hunting for game to supplement our rations. He returned dragging something through the long grass. I thought he might have shot a bird or even a young capybara with his bow and arrow. He had in fact killed a small spectacled caiman (*Caiman crocodilus*). I always said I would never eat crocodilian meat, as I am so fond of them. But it would have been boorish to refuse something Kennard had killed specifically for us. The spectacled caiman is in no danger and is sometimes hunted for food. Kennard had made a clean kill, shooting it through the skull with an iron-tipped arrow. It made you appreciate just how tough crocodilians are. The two-inch iron arrowhead had been bent right round by the caiman's hard hide and bone.

As the sun got hotter, we returned to camp. Jon and Chris decided to walk with Kennard to a ranch a few miles away to get some pop and water. We had all been drinking out of flasks that purified water from streams and ponds. The thought of pop was appealing, but I also thought they were both insane for wanting to walk in the blazing noon sun.

After they left, the heat began to rise. Even the guides said it was remarkably hot. The sun was truly unbearable. There was no respite in the tents, as they merely magnified the already savage temperature. We poured water over our heads, but that only elevated the heat for a short while. Finally, in desperation I waded into a swamp and stood under a tree for several hours. The mosquitoes were nothing, a minor irritation in comparison with the solar torture I was enduring. I have always preferred the warmth to the cold. I have always favoured summer over winter and sun over rain, but on that day I prayed for rain, or indeed the slightest cooling breeze. Now that I have felt the wrath of Guyana's dry season, I will never feel quite the same about cold weather.

Paul also began to suffer badly from the heat at this point. He later almost blacked out whilst packing up his tent.

Jon, Chris, and Kennard finally returned with pop and water. It was a nice change to be able to gulp down liquid rather than having to suck it up from a water-purifying flask.

That evening we moved on to Cashew Pond in an attempt to avoid the insects. The walk was supposed to have been 1.2 miles, but it was probably closer to 4 miles. The terrain was very bad. The uneven ground was dotted with concrete-hard lumps that were hidden by the dry grass. At Cashew Pond, Damon pointed out some cashew fruit. They grow directly below the nut. The nut is poisonous until roasted. In the West, we see only the nut and never the fruit. The bright yellow fruits looked a lot like pepper, but tasted like cranberry—juicy but with a paradoxically dry aftertaste.

That night we roasted the caiman on an open fire. The best meat was in the legs and tail. The legs tasted like chicken, whilst the meat of the tail tasted like very flavoursome, chewy cod. Once again, the insects made a meal out of me. In fact, we were all bitten worse here than at Crane Pond.

In the morning we packed up to move out to Point Ranch, where we were to be picked up by minibus and be taken back to Lethem. The walk was long, hot, and uncomfortable.

Whilst we waited at Point Ranch, I asked about the water tiger. An old man from the ranch named Elmo had seen them. He was adamant that they were not the giant otter (*Pteronura brasiliensis*) with which he was familiar. Elmo said that the water tigers he saw were spotted like a jaguar (*Panthera onca*) but hunted in a pack. He said that there was a "master," possibly a parent water tiger, that sent out the cubs ahead of it in order to flush out prey. He had seen a whole group of them several years ago. Elmo also said that a pack of water tigers lived on a local mountain. He pointed it out. It had no name, but it was said that a dragon guarded a spring on the same mountain. Elmo said that no one who had ever climbed it had returned.

Kennard confirmed that water tiger was supposed to come in different colours, spotted like a jaguar, brown, and white with dark spots.

Another of our guides, Joseph, stated that he had seen the hide of a water tiger killed by a hunter in the 1970s. It was ten feet long, including the long tail. It was white (he compared the shade with some cows on the ranch) and had black spots. The head was still attached. He said that it was striped like a tiger.

These descriptions, both physical and of behaviour, match no known cat species. We thought that the water tiger might actually be a form of giant mustelid, as certain species such as stoats can change the colour of their coats.

Joseph had another even stranger story to tell. In 1975, a plane crashed into a mountain in the same range as the one supposedly inhabited by a dragon and a family of water tigers. He was paid to climb up and retrieve the body of the pilot from the crash site. He found the wreckage and the corpse. It was missing its head and had been badly burnt. He retrieved it and put it into a sling fashioned from a blanket and began to descend. But on the way down he became hopelessly lost. He wandered for three days on the mountain before finding his way down. During this time, he was forced to consume the flesh of the crash victim. He ate the arm of the body in order to survive.

We took a truck back to Lethem. We had decided that none of us could stand another eighteen-plus-hour journey in a rickety bus, so we decided to fly back to Georgetown when we left. We booked our tickets in advance at the little airstrip in Lethem. Whilst there, we enquired about chartering a helicopter, but there was only one in the whole of Guyana and that was not available. Getting one from Brazil would have meant days of red tape. We considered a boat, but Kennard said that the river was too low. This was amazingly frustrating. The main thrust of the expedition was to search for a giant anaconda at Corona Falls, but we had not even seen one solitary ordinary-sized anaconda. The fact that the pool was unreachable was like a dangling carrot before a donkey. It looked like we would not be reaching our target on this trip. We decided to try and return another time during the rainy season and charter a boat or plane to get to Corona Falls. There was an airstrip only half an hour's walk from the pool, but no aircraft available presently.

Both Kennard and Damon mentioned that during the airstrip's construction a number of years ago, eleven skeletons had been found inside of termite mounds. The skeletons were in crouched positions, suggesting that people had broken open the termite mounds and placed the bodies inside them, leaving the termites to rebuild their mounds around the cadavers.

Neither Damon nor Kennard knew of any tradition in any Amerindian tribes that had funerary rites like this. Damon postulated that it could have been prehistoric. The bones had been thrown away and no further research done.

We checked into a guest house again. That night a heard of horses stampeded through the hotel grounds. None of the locals batted an eyelid! The next day Damon had arranged for us to meet with a former tribal chief who knew a lot about the strange creatures of Guyana.

We drove until the savannah changed to jungle, then we drove up a twisting jungle path to a clearing near a stream. Waiting for us was a middle-aged man in a Sideshow Bob t-shirt, holding a parang. He introduced himself as Ernest. He had been a tribal chief until about eight years before, when he retired to concentrate on running a little fish farm at the base of the Kanaku Mountains.

Ernest was a wealth of information regarding all of Guyana's monsters and then some. About ten years ago and around thirty miles away, he had seen a thirty-foot anaconda in a pool. He said that an Englishman had shot it and transported the skin to England. This, if it was imported, would have been done illegally. He knew of the di-di but had not seen one himself. However, a friend of his, who had died two years previously, had seen a di-di. He had seen a female suckling an infant in a tree. He had watched them for a while before blacking out. Afterwards he fell ill. His illness became worse and worse, and he blamed it on seeing the di-di. He only admitted to the sighting on his deathbed. Ernest said that the voice of the di-di, like a very loud human shout, was still heard from time to time in the Kanaku Mountains.

Ernest knew of the little red-faced pygmies. When he was nineteen (he was then fifty-nine) he had seen one. It was naked, brown-skinned and had a red face. Unlike Kennard, he felt that the red face was natural pigment and not painted on. The little man had taken tobacco from Ernest, then vanished back into the forest. He told us that the pygmies were more often seen than heard. They liked tobacco and were not dangerous unless angered. They sometimes made homes under large trees. If one of these were cut down, the pygmies, quite naturally, would get angry. They also made little pots that humans sometimes came across in the forest. These too should be left alone.

The pygmies did not speak to humans, even when spoken to. They seemed just to take tobacco and leave.

Damon confessed that he had seen a pygmy as well. About ten years ago, he had been camping with his sister-in-law and another girl. He awoke in his tent to see a tiny red-faced man grinning down at him. He was frozen with fear. Finally, he found he could move enough to try to nudge the girls awake. When he looked again, the pygmy had gone. He did not recall hearing the zip on his tent being pulled open.

At the age of twenty, Ernest had a run-in with the water tiger. He and his uncle were on a small boat on the river when something seized the vessel from beneath and started to shake it. Ernest and his uncle had to hold onto some branches overhanging the river in order to stop the boat from overturning. Ernest's uncle said it was a water tiger, though neither man saw the attacker. It could just as easily have been a big caiman. He too said that the water tiger lived in rivers and ran in packs.

His final story was of something that none of us had previously heard of. A couple of years before, in a little cave at a place called Wa-sa-roo, he had seen a tiny caiman. It was smaller even than the smallest known species, Cuvier's dwarf caiman (*Paleosuchus palpebrosus*). It was brown in colour and had a red strip running down its back. The description matches no known caiman species. But, stranger than this, he said that the tiny caiman had two tails!

Lizards and snakes have sometimes thrown up freak specimens with two tails due to genetic deformity. However, as far as I am aware, this has never been recorded in crocodilians. Could Ernest have seen a pair of caiman mating, one on top of the other? He did say that the tiny animal was making a very loud bellowing noise, out of proportion to its modest size. Male alligators are known to bellow loudly during mating season, so this seems like a reasonable explanation. Ernest had also seen the little caiman in a creek near the cave.

We thanked Ernest and set out to explore Wa-sa-roo. It was a collection of boulders of up to house size through which a stream ran. I took off my shoes and scrambled into the small cave. It was cool, had water and ledges. Though there was no evidence in the form of tracks, the cave was the perfect place for a small caiman to make its lair. Dwarf caimans like fairly fast-flowing, rocky streams. Guyana could be playing host to a new, unrecorded species.

We then climbed up onto the top of the boulders to look down into the caves. We used a night-vision camera to take film of the inside of the caves.

Later, back at the guest house, Kennard told us some more stories he had heard of the di-di. Once, many years earlier, a hunter found a huge human-like footprint miles from any habitation. He followed the tracks till they came to a tree. Looking up, he saw a huge hairy man sleeping in the vines. He ran away in fear.

In the 1940s, a girl was kidnapped by a di-di and taken deep into the jungle. It took her as its mate. And together they had a hybrid child, half human, half di-di. The girl stayed with the di-di against her will, until one day she saw a hunter in a canoe. She shouted him over to the bank and leapt aboard. As the hunter paddled off, the di-di emerged from the jungle and stood on the bank gesticulating for the girl to return. When she did not, the monster picked up their half-breed offspring and tore it to shreds like a doll.

Quite where this story had its genesis is an interesting question. I have heard exactly the same story told about both the yeti and the sasquatch.

Another story told of a group of men boating down the Essequeibo River. At one point they had stopped and disembarked. The men saw one of their colleagues grabbed from behind by a massive hairy arm. A huge ape-like figure carried him into the jungle. The other men pursued, shooting at the shaggy giant until it let the man drop.

That night we were invited to dine with Ernest's family at their home just outside of Lehem. During the meal Damon mentioned that several members of his family had once worked in the largest open-cast gold mine in Guyana. It was once owned by a Canadian company but has since been sold. They employed many Amerindians. One time a whole village full of people witnessed the uncovering of a huge, human-like skull. It was far larger than a man's skull, and Damon wondered if it could belong to *Gigantopithecus blacki*, a giant Asian ape believed extinct for 50,000 years. Many think that the yeti and sasquatch may be a surviving form of this ape or something related to it. So far *Gigantopithecus* remains have only been found in China and India. Officials from the company owning the mine came and took the skull, and it was never heard of again. Perhaps the company was scared of having the mine closed down if a major palaeontological discovery was made. Maybe this skull is locked in some bureaucrat's basement to this day.

Foster recalled an even odder story of some man-like creature with webbed digits being swept into a village during a flood. It was described as resembling "the Creature from the Black Lagoon." I too have a vague memory of hearing something like this many years ago. Just what the creature was, if it ever existed, and what had become of it, was unknown. Details were lacking, but Damon was sure it had appeared in the local paper many years ago. There is a tradition that is widespread in South America of small black aquatic goblins. There is little information on them, and whether they are based on some real creature remains to be seen.

The following day we caught the tiny plane back to Georgetown. Damon and Kennard had to stay in Lethem, as they had to pick up some snakes and invertebrates from remote villages.

Whilst waiting for the plane we met Rhiannon, the American girl we had seen before on the ferry. She told us that she had been researching a kind of spirit/creature called a Kanima. It seemed to be a different thing to each tribe. One said it took the form of a short fat man.

Foster said he had spotted a man who knew all about the Kanima and went over to fetch him. He was a rubber tapper by trade and fashioned his rubber into nice model birds that he sold to shops. He told us that the Kanima was a sort of solitary witch doctor, a human with magical powers. Kanima lived alone and would pass on their knowledge to suitable students who sought them out. Only men were Kanima. Kanima could lay curses or cast spells, but they usually left you alone unless you provoked them. One power he mentioned was to use certain leaves to become invisible in the forest. This may sound unbelievable, but it might be a matter of what is meant by "invisible." Tribesmen in Peru have long said that the secretions of the giant monkey tree frog (*Phyllomedusa bicolor*) made men invisible, gave them great stamina, and allowed them to go without food. Recently, biochemists analyzing the frog's secretions have found powerful laxatives, diuretics, and emetics that may well flush out smelly compounds from human skin, making the hunter "invisible" to forest animals that mainly rely on their sense of smell. The secretions seem to include painkillers and hunger suppressants in them as well. Could the plants used by the Kanimas have similar properties?

The rubber tapper also said he had seen red-faced pygmies as a boy. He had been hunting in the general area and saw a tiny man with a red face peering at him through the undergrowth. Unlike other witnesses, he thought the man was hairy (though this may have been an animal hide he was wearing). When he realized he had been seen, he fell to all fours and ran off. This is the only mention of hair and moving on all fours. Could this witness have mistaken a monkey? The red-faced black spider monkey (*Ateles paniscus*) fits the description. It is hairy and has a red face. It moves on all fours but can stand erect. However, it has a long tail and inhabits rain forest, not the dry savannah around Lethem.

Finally, we had to leave to catch the plane back to cold, dark, wet England.

So what are my final thoughts? Not getting to Corona Falls and the giant anaconda lair was a blow. We intend to rectify this in a future expedition. On the upside, we did turn up fascinating information on other cryptids.

The little red-faced men have never been recorded or written about anywhere else to my knowledge. They could be a type of tiny hominin, related perhaps to *Homo floresiensis*. They still seem to be about, but no one has ever studied them.

The di-di may be a bigger hominin, something related to the sasquatch perhaps. If a little foot can exist in Guyana, why not a bigfoot? Much of what is attached to them seems like universal folklore. They seem less common now than the pygmies. It seems whenever human habitation springs up, they retreat further into the wilderness.

The water tiger, despite my initial suspicions, seems to be very different from the giant otter (*Pteronura brasiliensis*). It is social, aggressive, and comes in several colour variations. It is a flesh-eating mammal of some kind, possibly a felid or mustelid.

Lastly, the tiny caiman. This is intriguing, and could constitute a whole new species. We need to gather more information and eyewitness accounts.

Guyana is a veritable menagerie of cryptids. Few expeditions have looked for these creatures before, so the country promises to be fertile ground for research for many years to come.

The Indian Yeti

The expeditionary team of Dr. Chris Clark, Adam Davies, Dave Archer, and myself, who had previously searched for the Russian almasty (a relic hominid) and the puzzling Sumatra orang-pendek (mystery ape or hominid), were getting our heads together in planning where to go in 2010.

Several years before, Adam had been in Tibet on the track of the yeti. Ian Redmond, a tropical field biologist and conservationist, mentioned to him that there were numerous reports of the yeti in the northern Indian state of Meghalaya. Upon returning to England, he investigated more closely and found that local documentary filmmaker and journalist Dipu Marak had been on the trail of the creature for some years.

I too had heard of the Indian yeti, or as it is locally known, "mande-ba-rung," the forest man. In June of 2008, BBC journalist Alistair Lawson visited the area to investigate sightings of the creature. He was impressed by the remote, undisturbed landscape and wrote:

> *If ever there was terrain where a peace-loving yeti could live its life undisturbed by human interference, then this has surely got to be it. Perhaps the most famous reported sighting was in April 2002, when forestry officer James Marak was among a team of fourteen officials carrying out a census of tigers in Balpakram when they saw what they thought was a yeti.*

Dipu had given the BBC some hairs he had found at a remote area called Balpakram. Upon analysis, these proved to be from a goral (*Nemorhaedus goral*), a species of Asian wild goat. This, however, did not negate the eye-witness reports.

We decided that the CFZ team should investigate and began to lay plans for a trip to India. Adam, who is a great organizer, contacted Dipu, who in turn organized guides, lodges, and contact with eyewitnesses.

The four of us were to be joined on this trip by Jonathan McGowen. Jon is an excellent field naturalist and taxidermist, as well as the curator of the Bournemouth Natural History Museum.

On Halloween 2010, we flew out to India. During the long journey, Chris collapsed and we called for a doctor. He was given oxygen and quickly recovered. The verdict was that he had been suffering from a lack of oxygen on the long stuffy flight.

We arrived in the mad cacophony that is Delhi in the evening and checked into our hotel. We had an evening to kill, so we arranged for a taxi to show us some of the sights of the city. Unfortunately, the taxi driver just dumped us at a Western-style mall, thinking that, as Westerners, it would be the place we were most interested in!

Finally, we made it back to the hotel, after avoiding wandering cows in the road and a near-collision with a surprised Sikh in a three-wheeler. Driving in Asian cities is certainly an experience. It seems to consist of 90 percent horn beeping. In India there is even a horn code, certain numbers of beeps meaning certain things. The legend "Horn Please" is amusingly written on the back of many vehicles.

The following day we flew out from the surprisingly clean and efficient Delhi airport to Guwahati in Assam. We were met at Guwahati by our chief guide Rudy Sangma, assistant guide Pintu, and our drivers. We then began the long journey to the town of Tura in the West Garo Hills.

Meghalaya is a mountainous state in the northeast of India. It was carved out of the state of Assam in 1972 to accommodate the Khasi, Garo, and Jaintia tribes, who at one time each had their own kingdoms. The three territories had come under British administration in the early 1800s and were assimilated into Assam in 1835. Once fierce head-hunters, the Garos were among the first Indians to be converted to Christianity by British missionaries. After conversion, the tribes were largely left alone, allowing a lot of their culture to remain intact.

This expedition was to be somewhat atypical. Generally, we camp out in the jungle, mountains, desert, or wherever twenty-four-seven, returning only to civilization to stock up on supplies. This time, however, the Indian Government would not let us stay overnight in the jungle, due to the activities of the insurgent group the Garo National Liberation Army. The significantly reduced our chances of seeing the mande-barung.

As the winding roads rose upwards, giving way to rocky tracks, Rudy told me of some of the other strange creatures from the folklore of the Garo Hills. One creature that looms large in the Garos is the sankuni. This is a monstrous snake that bears a crest upon its head much like rooster's comb. The description of the sankuni matches up very well with that of the naga, the vast crested serpent I searched for in Thailand back in 2000, and the ninki-nanka, the serpent dragon of the Gambia I hunted in 2006. All are said to bear crests, be of huge size, have shining black scales, live in lakes or rivers as well as subterranean burrows, and have an association with rain. The uncanny dovetailing of these stories made we seriously wonder if the sankuni and other monster snakes are based on encounters with a real-life species of gigantic snake unknown to science. Unlike the naga or ninki-nanka, the sankuni is also associated with landslides. Its underground crawling is supposed to cause massive shifts in wet earth. This sounds much like the weird South American serpentine cryptid known as the minhoco that is said to cause disruption, uproot trees, destroy houses, and even alter the course of rivers. The sankuni is not wholly malevolent. Indeed, in legend, it is said to allow humans to use its great coils as a bridge, allowing them to cross rivers. It is also said to manifest in dreams, warning people of impending landslides. The sankuni is said to crow like a rooster, much the same as the crested crown cobra of Africa. Its likeness to both the European basilisk (save for its vast size) and the giant serpentine lindorms and worms hardly needs to be stated.

Another weird entity from folklore that Rudy told me of was the skaul. This is a vampiric entity that resembles a normal human being by day, but at night its head detaches from its body and flies about as an independent entity. It has luminous hair and saliva. The skaul is said to feed on excrement and rubbish, but also to suck up human life force, causing the victim to fall ill, weaken, and finally die. The skaul may have been an early attempt to explain disease and illness. The luminous hair and saliva might well be based on early sightings of ball lightning or some other metrological phenomenon. The skaul has analogues across Asia with the Malayan Penanggalan, the Philippine Manananggal, the Balinese Leyak, the Thai Krasue, and the Japanese Nukekubi.

We arrived in the ugly mountain town of Tura. Soulless grey buildings sprung up like crowded fungus whilst gas and water pipes snaked above ground like rusting metal mycelia. The town was dirty, noisy, and smelly. We checked into the tumbledown Sandre Hotel and unpacked.

Tura's unappealing nature was offset by our meeting with Dipu Marak, the man who had been on the trail of the mande-barung for many years. A delightful man, Dipu has a deep and infectious passion for the Indian yeti. He told us how he recalled hearing stories of the beast in his childhood and that sparked his lifelong interest. With a Garo mother and a Bengali father, Dipu is a huge fellow who towers over everyone else in the town.

The native Garos are quite distinct from the average Indian. They are a race that originated in Tibet, fought their way down to India, and finally settled in the hills that bear their name to this day. Property and land are passed down the female side of the family, a wise move in a people who had to fight every step of the way on their long migration.

The following day we journeyed to Nokrek National Park. The hills here are covered by deep virgin rain forest. It was here that we intended to leave our camera traps for the duration of the expedition. It was here that the wild ancestor of all modern varieties of orange was discovered. I sampled some *Citrus indica*, finding it to taste like a less-sharp lemon. Another plant growing in abundance was a small, dirty brown, spherical fruit the locals called "tastecan." They looked like oversized oak gall, but tasted exquisite. The flavour was quite unlike anything I have ever tried before. To attempt to describe it would be akin to trying to describe a new colour. Rudy told us that a few weeks before, the area was swarming with elephants and wild buffalo, but they had now moved on. We heard hoolock gibbons calling in the distance.

We took an arduous trek into the rain forest. The terrain was very hilly and we were constantly climbing up and down ridges. We came across the nest of a wild boar and climbed down a dangerously steep cliff to investigate a small cave. The cave offered up no results other than the paw prints of a small felid, possibly an Indian leopard cat.

We planted camera traps at several locations, making sure each had a good view of the area. All the traps were baited with bananas and oranges.

The mande-barung is supposed to be primarily an herbivore, although there are a couple of sightings of the creatures eating freshwater crabs. Dipu told us of one case where a farmer saw a family of four mande-barung stealing pineapples from his fields. The creatures ran away upon seeing him, snatching fruit as they went.

We moved from Tura in the West Garos down to Siju in the South Garos. We were met by Rufus, a friend of Rudy's and another guide. We stopped in a rather down-at-heel and basic, but clean, tourist lodge. Close by were the Siju Caves, where the village head man had supposedly encountered a mande-barung several years before. The whole area was awash with wildlife, from Indian false vampire bats and tokay geckoes in the kitchen to tarantulas in the walls outside. Jon McGowen used some fishing line and a live cricket to go tarantula fishing, baiting the spider out far enough to be photographed.

The caves themselves were amazing. Apparently, they go on for miles, with many smaller passages branching off the main cave. Fulvous fruit bats (*Rousettus leschenaultia*) roosted in the cave and bizarre white fungus sprouted up from their droppings. The waters that ran through the cave were alive with tiny fish, shrimp, crab, and cave crayfish. A swarm of them were feeding on a dead bat. Jon found two recently dead bats and decided to take them back with him to be stuffed. Huntsman spiders as broad as a human hand scurried over the rocks. I was excavating in the earth of the cave in the hope of finding some bone material. Pintu, one of the guides/porters found a section of what looked like leg bone under some rocks. It was around six inches long. Upon examining it in the daylight, Jon thought it looked like the femur of a biped. We kept it for analysis.

The following day we set out across a huge suspension bridge that spanned the Simsang River and began to trek into the jungle. Early on, Chris complained of pains in his chest and turned back, leaving us all quite worried. Despite being the eldest among us, on previous expeditions he had romped up mountains, across deserts, and through jungles that left the rest of us gasping. It was clear that there was something wrong with him.

As we entered the jungle, a huge Bengal eagle owl went crashing through the canopy. As the path rose, we glimpsed wild jungle fowl, ancestor of the domestic chicken. This place really did remind me of Kipling's India. We came across an area of limestone outcroppings in the jungle. Some had been sculpted by wind and water to resemble human faces; other looked like the walls of lost temples or ruined cities, though all were natural in formation. They brought to mind the Cold Lairs where the *Bandar-Log*, or monkey people, brought Mowgli in *The Jungle Book*. One formation in particular was a narrow passage between two limestone cliffs. Rudy and Rufus told us that, up until around twenty years ago, the passage was used by hunting tigers to ambush men. Humans were forced to walk single file, and the walls were too steep and slippery to climb, making the men easy prey for the great cats. Later, we came upon a watering hole and searched the mud for tracks; we found elephant, sambur, barking deer, and buffalo. At one point, as we were resting in the jungle, something leapt down from the trees just over a ridge above us. The guides thought it may have been a leopard that was stalking us, but on examination they said it was more likely to have been a monkey. Indeed, though we saw none, we did find many monkey droppings.

The paradox of the jungle is that, although it contains the greatest concentration of life anywhere on earth, animals are more difficult to see here than anywhere else. Creatures can hear a human coming from a long way off and melt like ghosts into the shadows. Wildlife is much easier to spot in open grassland areas. In all my time in many rain forests around the world, I've only seen a handful of large animals.

Whilst most of us had been away in the jungle, Dave Archer had stayed by the Simsang, searching for snakes and looking for animal tracks. He had found the footprints of a tigress in the sand. It was good to know that there were still tigers in the area.

Later we interviewed the head man of the village, Gentar. He had encountered something strange in the Siju Caves several years before, something that had frightened him so much that he refused to go back there. He and some friends had been fishing by the light of burning torches. They had heard a noise that he described as sounding like someone treading on bamboo. On investigation, they found wet footprints on the rocks. They were human-like but of a vast size. They led down one of the passages that turned off the main one. The group thought a mande-barung had entered the cave from one of its many jungle entrances. They panicked and fled the cave.

I found it odd that such a creature would be lurking so close to human habitation, but I was to hear subsequent stories of them approaching other villages. Cave systems retain a stable temperature; it could be that the creature had entered the caves to keep cool, or possibly to hunt for crabs.

Rudy and Rufus told us of their worries over the future of Garo culture. The younger generation are losing interest and increasingly wanting to become Westernized. Only the very remote tribes are still animist and still hold onto all the old beliefs that are beginning to die out elsewhere. They are planning to write a book recording Garo culture and custom before it is lost.

From Siju we moved down to Bagimara and set up HQ in a delightful lodge with a magnificent view of the Simsang. In the evening, we would watch the sun setting over the river from the veranda. I enjoyed several chapters of Kipling's immortal Jungle Book series, so cheapened and bastardized by Disney. Of all the places we stopped in India, this was my favourite.

Whilst here, we were introduced to a local man called Beka. A sculptor by trade, he had an interest in cryptozoology. He told us of a story his father related to him. Around 1940, in a lake near the border of Bangladesh, a group of armed men, possibly soldiers, had shot a sankuni. Apparently, the creature had devoured a number of people over the years. The creature's body lay partly out of the lake and partly in. The portion out of the lake was said to measure sixty feet. If there was any truth to the story, it made me wonder just what kind of firearms would be needed to do any kind of serious damage to a snake so huge, and what happened to the body? The story might be nothing more than a tall tale, but it highlights the belief in a giant crested serpent in the Garos.

More recently, within the last five years, there had been a case of a woman who dreamed that a man had warned her that her house was going to be destroyed due to an impeding landslide. She moved out of her house, and it was indeed destroyed by a landslide. Witnesses saw a huge sankuni crawling away from the wreckage. It could be, that if the sankuni is a real flesh-and-blood animal, it inhabits underground burrows and lairs. If these are disturbed by a landslide and the animal is seen crawling away, then people may have thought that the sankuni's coils had been the cause of the landslide.

We travelled to the village of Imangri and trekked into the jungle beyond. We saw a simulacrum of a footprint in limestone beside a river. It is a natural formation, but the fact it has been linked with the mande-burung argues that the creatures must have been known of for a long time for such an association to have arisen. Swarms of yellow butterflies flitted around, and we rested awhile beside the waters. Chris once more felt ill and stayed behind in the village.

We returned to Imangri and interviewed the head man, Shireng R. Marak, a fifty-six-year-old with two thumbs on his right hand. In 1978, he and some friends were hunting in the forest. As it was beginning to grow dark, he heard something big and powerful crashing through the forest. He heard a loud, deep call, *AUHH!-AUHH-AUHH!* which he imitated for us loudly. He had heard village elders talking about the mande-barung and demonstrating the sound it made. He and his friends ran into a cave and lit a fire at the entrance. They heard the creature bellowing and crashing around outside the cave all night. At first light, it moved away into the forest, and they ran back to the village.

Shireng said that sightings of the creature were more common forty years ago. His friend's grandfather had shot one. He said it was man-like, covered in black fur, with a face like a monkey.

The village shaman had supposedly seen the mande-barung as well. He was out in the fields somewhere, so a boy from the village was sent to look for him. In the meantime, I took a short trip by canoe down the Simsang River. Finally, the shaman was found, and we interviewed him over tea.

Neka Marak was seventy-seven and was now suffering from cataracts. He made medicines and charms. Back before the Indo-Pakistan War of 1965, he had been searching for an incense tree in the jungle. He came upon some thick creepers that had been snapped by something with immense strength. He heard a crashing sound and turned around to see a huge mande-barung charging at him through the jungle. Neka pointed to the roof of a nearby tea house in the village in order to give us an idea of the size of the creature. The roof of the tea house was fifteen feet high, a size I was totally unable to accept for the mande-barung. I don't know if it was the old man's cataracts making him overestimate, or the length of time since the sighting, or sheer fear. He went on to say it resembled a huge hair-covered man. The face looked very human, and the hands were big enough to have broken a human's neck. After all this time, he could not recall the colour of the creature's fur. To his credit, he did not try to embellish, but he admitted that he could not recall the colour of the hair. He fled from the forest as quickly as he could.

Neka had also seen the sankuni. This also occurred prior to 1965. He saw the creature emerge from a cave beside the Simsang River. He did not see the whole animal, as he beat a hasty retreat. He indicated that the portion he saw was in the region of twenty-five to thirty feet long. It was black-scaled with a yellowish underbelly. It had a red, rooster-like crest and red wattles under the lower jaw. He fled in terror from the giant snake.

The following day we drove to Balpakram, an area that looms high in Garo legend. It was thought to be the place where the souls of the dead rested before going into the next world. It is a national park, and the forested areas are full of wildlife.

The roads grew more treacherous as we drove higher. Soon even the four-wheel-drives were struggling to cope. We walked the final couple of miles on foot to the great plateau that formed Balpakram. I noticed that the area was heavily used for grazing, and there were quite a few people around. Herdsmen were burning off the dry grass to promote new growth for grazing their livestock. I found it hard to visualize a large ape, or indeed any big animal, existing in the area. The basalt rocks in the park were in six-sided geometric forms much like those in the Giant's Causeway in Ireland. The molten rock formed the shapes as it cooled and contracted. Unlike in Ireland, there are no columns, and the shapes are visible only at ground level. Local people call the strange configuration the ghost market. Rudy told us that fossil pumpkin, melon, and tomato seeds have been found in the area, leading to the legend that it is a place where spirits hold a marketplace at night.

As we walked further across the plateau, we finally came across a truly spectacular gorge. Seven kilometres wide, two kilometres across, and around one kilometre deep, the Balpakram gorge made an astounding spectacle. It was heavily forested and had near-sheer sides. A river ran through the bottom, and Rudy explained that the only safe way in was via canoe by the river from a nearby village. Only two or so hunters ventured into the gorge per year, and it was mostly unexplored. It looked as if it could easily hide a small group of yeti in its deep, inaccessible forests. Unfortunately, we did not have enough time to investigate the gorge, as such an undertaking would have taken a whole week. We made plans to return to the gorge on a future expedition.

Back at the lodge, we met owner Bullbully Marak, who told us how keen she was to promote eco-tourism in the area. The Garo Hills and Meghalaya in general are not often visited by tourists. Rudy and Rufus mentioned that they often feel like foreigners in their own country and are often mistaken for Indonesians or Malayans. The feeling throughout the Garos is that the central government of India is ignoring them. Such feelings have led to the formation of several insurgent groups in the area.

We returned to the somewhat depressing surroundings of Tura and met up with Dipu once more. For approximately one year, the CFZ had been involved with the making of a major documentary for the BBC. Minnow Films had already covered our annual convention, the Weird Weekend, and had briefly followed CFZ Director Jon Downes and his wife Corinna to Texas, where he was investigating blue-skinned hairless dogs that seemed to be hybrids of the coyote and the ultra-rare Mexican wolf.

Cameraman, BAFTA-winning director, and company founder Morgan Mathews and the strikingly beautiful soundwoman Tara Nolan had made an epic journey to join us. Despite not having slept in nearly two days, they were keen to begin filming. There had been some initial worry that having a film crew, even a tiny one, would get in the way of the expedition. As it turned out, they were fine and never once slowed us down or in any way interfered with the project.

Tura itself is devoid of anything approaching nightlife. The one bar in the town was at the Sandre Hotel and closed at ten o'clock sharp. The bartender seemed totally disinterested in making money and resented anyone who entered the bar after 9:45.

The two hotels in Tura both had restaurants that were spectacularly badly run. Their menus were surprisingly varied, but most items on them were not available. This made ordering food a bit like the cheese shop sketch by Monty Python. Far worse than this was the service. On one occasion, we ordered some soft drinks. An hour later they still had not arrived, despite three waiters standing around next to the fridge that the drinks were in. Dipu himself had to go and open the fridge and point the drinks out to them. On another occasion, I ordered soup and bread. The soup took an hour to come and the bread turned up an hour later.

We were to spend the next day interviewing a number of people around Tura. The first on our list was Dr. Milton Sasama, the Pro-Vice-Chancellor of Garo Hills University. He had written a number of books on the history and folklore of the Garo Hills. He did not believe in the mande-burung, as he had never come across descriptions of the beast in any of his studies. He had only heard of the monster, like a giant orang-utan, in the past twenty years. He also asserted that there was no tradition of a yeti-like creature in Assam, the Indian state that lies between the Garo Hills and the Himalayas.

Conversely, he believed implicitly in the sankuni. He knew a man who had eaten the flesh of a dead, juvenile sankuni after it had been washed into a village by a flood. It was between twelve and twenty feet long and bore a rooster-like crest. The meat from the carcass had provided enough food for the whole village. The man, now in his eighties, called Albin Stone, resided in Tura these days.

Our next interview was with Llewellyn Marak, the uncle of Rufus, who was a noted naturalist and author of a number of books on the wildlife of the Garo Hills. In 1999, he came across a set of four huge, man-like footprints at Nokrek Peak, around twenty-one kilometres from Tura. They were found in sand beside a stream and were eighteen inches long. The tracks led away into the jungle.

Llewellyn's grandfather was a renowned hunter who amassed a large collection of trophies. He had encountered the mande-barung on a hunting trip many years before. He said that he came across the beast in a jungle clearing. It resembled a huge gorilla and was black in colour. It moved around on all fours and seemed to be searching for food. Occasionally it would stop and sit, appearing to eat something. Llewellyn's grandfather became afraid and backed away.

This is the only report we have of the creature moving on all fours. Then again, it may have been doing this in order to forage for food. The experienced hunter was sure what he had seen was not a bear.

Llewellyn, a conservationist rather than a hunter, invited us to look at his father's collection. Eagle-eyed Jon McGowan spotted something unusual among them. There was a pair of muntjac horns of unbelievable size.

On closer examination, these very distinctive horns proved to be even larger than those of the giant muntjac *(Muntiacus vuquangensis)* of Vietnam and Laos. The accompanying photo shows the horns next to those of the Indian muntjac *(Muntiacus muntjak)*; the startling size difference is apparent. We took some samples from the antler for analysis back in Europe.

Llewellyn had also heard stories of giant catfish and giant freshwater stingrays, much like those said to lurk in the Mekong River of Indo-China.

Following this, we moved on to speak to Rufus's uncle, a surgeon called Dr. Lao. Dr. Lao also believed that the mande-barung existed, but he thought that it was now very rare. Dr. Lao had a collection of books on Indian wildlife. Among them was a book entitled *A Naturalist in Karbi Anglong*, by Awaruddin Choudry, first published in 1993. The book, by one of India's best-known naturalists, records his time in the Karbi Anglong district of Assam, the Indian state to the north of Meghalaya.

One chapter of Choudry's book is given over to the Khenglong-po, a yeti-like creature seen in the area. As Assam borders onto Bhutan there is a link or corridor, if you will, directly from the Himalayas down to the Garo Hills, along which yetis are reported, which totally refutes Dr. Milton Sasama's assertion that no such creatures are reported from Assam. Choudry writes:

> *Singhason peak and some nearby areas are sacred to the Karbis.*
> *Here in the dense forest lives the Khenglong-po, the legendary*
> *"hairy wild-man." The Khenglong-po is an important figure in the*
> *Karbi folk tale. Whenever I used to get reports of its existence, I*
> *dismissed them as fable or mistaken identification of an ordinary*
> *animal. But when the much experienced Sarsing Rongphar gave*
> *me a fresh report, I had to re-think. Sarsing had been my guide*

in parts of the Dhansiri Reserved Forest, and I found him to be an accurate and reliable observer.

Sarsing was a hunter who used dogs to sniff out game, such as muntjac and porcupine, that he then dispatched with a long hunting knife. Even before his arrival at Karbi Along Awaruddin, Choudry had heard of sightings of a large, bipedal ape. At first, he asked witnesses if they might be mistaking a stump-tailed macaque (*Macaca arctoides*) or a hoolock gibbon (*Hoolock hoolock*), but the witnesses rejected this, as they were familiar with both species. But when his trusted guide told him of an encounter with the beast, Choudry was forced to change his mind.

It was on May 13, 1992, that Sarsing Rongphar and his friend Buraso Terang and his hunting dogs ventured into Dhansiri Reserved Forest. In the afternoon they came upon large man-like footprints that were around eighteen inches long and six to seven inches wide. The pair followed the tracks for three kilometers, until their usually brave dogs began to panic. Fearing an elephant or tiger was close by, they crept cautiously forward. Soon a loud breathing sound became audible, a "khhr-khhhr" sound. From 260–295 feet away, they saw an ape-like creature leaning against a tree, apparently asleep. The witnesses were at a higher elevation than the creature and had a clear view since there was no dense undergrowth to obscure it. The creature was jet-black, like a male hoolock gibbon, with thick bear-like hair on the body. The hair on the head was long and curly. The creature was a female with visible breasts. Its mouth was open and large, human-like teeth apparent. The face, hands, and feet were black and ape-like. In front of the creature was a broken tree, and the hunters thought the creature had been feeding on it. They observed the sleeping animal for around one hour. Sarsing likened it to a giant hoolock gibbon, but with much shorter forearms.

On reaching their village, they told tribal elders of what they had seen, and were informed that it was a Khenglong-po, a kind of hairy wild-man that was thought to be dangerous.

Choudry took Sarsing to his camp and showed him pictures of the Asian black bear (*Ursus thibetanus*) standing on its hind legs and the mountain gorilla (*Gorilla beringei beringei*). The hunter identified the latter creature as being a Khenglong-po whilst recognizing the former for exactly what it was. Choudry interviewed Buraso Terang separately and got the same answers.

A Khenglong-po was once supposed to have wandered up the railway track from Langcholiet to Nailalung.

On another occasion, Choudry talked to some hunters from Karbi Anglong in central Assam. They spoke of a large, herbivorous, ground-dwelling ape that they called Gammi. According to them, two Gammis were seen together in 1982, feeding on reeds on the eastern slope of the Karbi Plateau in the upper Deopani area. An elderly hunter had encountered one in the Intanki Reserved Forest in Nagaland in 1977–78. The creatures are said to be covered in grey hair and to be man-like in appearance. The name Gammi means "wild-man."

Choudry concludes:

> It seems possible to me that a terrestrial ape, larger than the gibbons, existed in some remote parts of Karbi Anglong and adjacent areas of Nagaland. The creature was always rare and preferred the remotest corner of the jungle, and, hence, evaded discovery by the scientific world. Now with the forests vanishing everywhere, this ape perhaps faces extinction. Expeditions to the heart of the Dhansiri Reserved Forest and Singhason area may well produce some result. But for now, I am looking for any fossil evidence including skull, bone, or part thereof. This will at least put the Khenglong-po at its right place, even if it is extinct. Lastly, if a large mammal like the Javan or smaller one-horned rhinoceros (Rhinoceros sondaicus) can be discovered in recent years in a small pocket of the war-ravaged Vietnam, outside its known locality in Indonesia and beyond anybody's expectation, one cannot rule out a Khenglong-po in the forests of Karbi Anglong.

We can see then an unbroken link of yeti sightings from Bhutan down into India.

The following day we interviewed another witness. He was a fifty-one-year-old teacher called Kingston. In 1987 he and a friend were on Tura Peak.

He saw large, five-toed, man-like tracks in wet sand beside a stream. The toes and heel extended far beyond his own. My size nines were bigger than Kingston's; he told me the creature's tracks were bigger than my feet. The tracks sunk an inch into the sand, whereas Kingston's own tracks only sunk in half an inch. He heard the mande-barung's cry, *AUHH!-AUHH-AUHH!* He imitated the sound, which was in line with that made by other witnesses. He wanted to investigate further, but his friend was too afraid. He said he has heard the cry on Tura Peak again, within the last few years.

We took tea with Kingston, and he told us of been bitten by a viper—from his description it was a white-lipped pit viper (*Trimeresurus albolabris*)— and how he had used a snakestone to draw out the venom and save his life. The use of snakestones is widespread in Asia, Africa, and South America. They are not really stones at all, but parts of animal bone that have been cut and shaped with sandpaper before being wrapped in foil and placed in a charcoal fire for fifteen to twenty minutes. The porous bone is said to draw out the snake's venom. Kingston said that the snakestone had adhered to the bite and "drew up" all the venom from his arm before falling off. Studies have shown that snakestones are nothing but a placebo peddled by quacks. Maybe Kingston was just very lucky and recovered naturally, or perhaps the viper did not inject the full payload of its venom. He told us the chemist in Tura still sold them. We thought it might be nice to buy some as souvenirs, but as it turned out snakestones were not to be found among the modern medicines on sale at the chemist.

Later that day, we visited the village of Apertee, some thirty-five miles from Tura, to meet a witness called Nicholas Sama. In the 1960s he had seen the severed hand and forearm of a mande-barung at a village market. The forearm, which was being displayed on a store selling bushmeat, was as long as his whole arm. The hand looked like a man's but far larger, and the nails were long. The arm was covered in long black hair. Nicholas thought it was very old, as the skin was desiccated. No one knew where it had originally come from. Nicholas knew what he was looking at was not the arm of a bear or a gibbon.

The next day we met with a most impressive witness in the village of Ronbakgre. Teng Sangma had heard that, in April of 2004, a village carpenter had seen a female mande-barung suckling an infant in a bamboo forest close to Rongarre. He did not believe the story, but then on the twenty-fourth of that month, he and a friend were hunting for jungle fowl in the forest when they came across a huge figure sitting with its back to them. Even in its sitting position, it was five feet tall. It was covered with dark hair and had longer hair on its head that fell down onto the shoulders and the back. The shoulders were very broad. It was a female and was suckling a youngster whose legs were visible at the side of its mother, suggesting that the infant was sitting on her lap. The youngster was making gurgling noises. The adult was pulling down large bamboo stems and plucking off the leaves to eat them. The men got to within fifty feet of the creatures and watched them for two minutes before they became afraid and backed away, leaving the creatures behind. Apparently, the creatures had not noticed them.

We explored the area, walking along a stream into the jungle. We found the tracks of a fishing cat (*Prionailurus viverrinus*) and what looked like barefoot human tracks. The latter were far too small to belong to an adult mande-barung, but we tried to take casts of them just in case. Unfortunately, the ground on which they were imprinted was far too damp to make casting with plaster of Paris possible. Prior to our leaving the UK, I had looked into other mediums for making casts. I could find no resin and was told by a DIY shop that Polyfilla was unsuitable. The only liquid rubber I could source came with a big, gun-like applicator that would be difficult to get through customs. We all agreed that the tracks were probably made by a human, but we filmed and photographed them anyhow.

The following day, we attended the Wangala Festival on the outskirts of Tura. The festival, also called the One Hundred Drums Festival, has its genesis in the pre-Christian tribal celebrations of the area. It was held in each village after the harvest. It is a "Thanksgiving" ceremony to Misi Saljong, also known as Pattigipa Ra'rongipa (the Great Giver) for having blessed the people with rich harvest of the season.

A day ahead of the Wangala, a ritual called the Rugala is performed by the Nokma (a village chief), and in this ritual, the offerings of special rice beer along with cooked rice and vegetables are made to Misi Saljong, the Giver. On the next day, the Nokma performs Cha'chat So'a ceremony, or the burning of incense, at the central pillar of his house to mark the beginning of the weeklong Wangala Festival.

With the influence of Christianity, the festival began to die out in all but the most remote villages. So in order to protect, preserve, and promote cultural identity, a group of Garos decided to organize the Wangala Festival on modern lines. A group of thirty dancers with ten drums would form a contingent, and three hundred dancers would make up the Hundred Drums Wangala Festival. It has been held every year since 1976.

We were lucky enough to be guests at this gathering and met a number of local dignitaries and tried some of the locally-brewed rice beer. The performers, who had come from all over the Garos, were brightly dressed in differing colours for each tribe. They had ten drummers apiece and dancing girls, as well as a warrior with a sword and wooden shield.

Each drum rhythm and dance was distinct and represented different aspects of life, the most memorable being a dance to represent "the shooing away of flies that are perching upon rice."

At the festival we noticed some other Westerners. They turned out to be French girls who were studying the hoolock gibbons. They introduced themselves and said that they had found our camera traps at Nokrek and apologized in case they turned up on any of the pictures. They had even written a letter to us and left it at the lodge, not realizing that they would meet us at the Wangala Festival.

Later we all had dinner at a tourist lodge. The lady who ran the lodge had recently cooked for the Prime Minister of India, and I could well believe it. "Dinner" is a term that does what we ate a disservice. It was a feast of positively medieval proportions, with whole chickens, huge fish, sides of pork, and masses of fruit.

The following day, we met another impressive witness. Nelbison Sangma was a farmer from the village of Sansasico. He observed a mande-barung for three days running in 2003. He was some 1,640 feet from the creature, looking down upon it. The creature was on the top of a smaller hill. When he first saw it, the mande-barung was standing under a tree. It was nine feet tall and covered with black hair. It moved around for an hour as he watched it. It then slept in a nest it had constructed by pulling down branches, much as a gorilla does. The next day, the creature was in the same place and appeared to be sunning itself. This time he watched it for half an hour. On the third day he saw it again, and it was wandering about and foraging.

The following day, he took some other villagers to the area and showed them the nest. There was a monkey-like smell that pervaded the surroundings. They found man-like tracks eighteen inches long and a huge dropping the length of a human forearm. This contained fibres from banana leaves.

We switched our attention back to Nokrek National Park. On the way, we picked up some provisions, rice, fruit, and several live chickens. These I named Little Lofty, Gloria, and Mr. La-di-dah Gunner Graham after characters off *It Ain't Half Hot Mum*. Dave bought a pot of the rice beer we had tried at the Wangala Festival from a roadside vendor.

We stayed in a specially-made tourist lodge near a village in the park. It was made to look like a traditional Garo house constructed of wood and bamboo. During the day's exploration, we came across a huge man-like track imprinted deep into the sand beside a stream. The print was like a human track but with a couple of important differences. The heel was proportionally broader, indicating a weight-bearing heel. The toes were more even than a human's, showing much less of curve from the big toe down to the little toe. The track was sunk over an inch into the wet sand, whereas my own footprints could not reach even half that depth. We photographed the print with frames of reference.

Back at the lodge, we ate dinner, then sat around the fire telling stories. I tried some of the rice beer Dave had brought. I was to regret it later. During the night I had severe stomach pains, as if someone were twisting a knife in my guts. I felt bloated and feverish. During one of my many visits to the toilet during the night, I heard something large moving around in the darkness outside the lodge. I assumed that it was a goat or a cow that had wandered down from the village.

In the morning, Rudy told us that something had been pushing against the lodge door. He thought it was a tiger that had been attracted by the live chickens we were keeping inside the lodge.

I was feeling worse than ever and was in a lot of pain. I forced myself to go with the others into the jungle on the off-chance that the creature would put in an appearance. I found the trek hard with my intense stomach pain and fever. At a waterhole I had to stop and rest as the others went on. Dave and Jon found another set of mande-barung tracks by a stream further into the jungle. They followed the stream and seemed fairly fresh. The creature seemed to be overturning rocks and hunting for freshwater crabs. Several crabs were discovered with their insides sucked out.

I managed to stagger back to the camp. Thankfully, by morning I was feeling better.

We returned to Tura, as Morgan and Tara had to leave for Delhi the next day. The next day was spent around Tura chasing up some leads and loose ends. We photocopied the relevant chapter from Dr. Lao's book, then visited the rather shabby library to see if there was anything on the mande-barung or the sankuni in any of the books there. We turned up absolutely nothing. We tried to track down Albin Stone, the man who was said to have eaten the flesh of a dead juvenile sankuni, but he was not at home.

Dipu showed us a rib bone found by his father at Balpakram in 1989. I thought it looked more bovid than primate, but we took a sample from it. Dipu also showed us a collection of hairs he found at Nokrek in 2006. The looked to me like goral (*Naemorhedus goral*), a goat-like antelope known to inhabit the area.

We met Dipu's uncle Garfield, who whilst fishing in 1956–57 came across a mande-barung print beside a stream. It was on a rock and formed by water where the creature had recently walked out of the stream and across the rocks.

Garfield also claimed to have seen the trail of a sankuni in the 1970s. It emerged from the Garo River and ran for three hundred feet under a wooden bridge, destroying some of the supports. It crossed a paddy field and entered a marsh. However, upon closer questioning, Garfield told us that there were two tracks running parallel to each other, and the ground and the vegetation between them was undisturbed. This sounds very like the tracks of some kind of all-terrain vehicle, rather than a giant snake that would leave one single furrow.

All too soon, our time in the Garos was over. We had to say our goodbyes to Dipu, Rudy, Rufus, and the others. We returned to Delhi, where we had a day to see the local sights.

The samples were tested by Lars Thomas and his team. The bone from the cave was a modern human and the antler from a known species of deer.

I am convinced that the mande-barung exists and that it is one and the same as the larger kind of yeti. The best model we have for this animal is a surviving form of *Gigantopithecus blacki*. As for the sankuni, its startling resemblance to the Indo-Chinese naga, the West African ninki-nanka, the Central African crested crowing cobra, and many other monster serpents, convinces me that there is more to these stories than hot air.

Already there is talk of returning to the Garos in a few years' time, probably to mount an expedition down into the gorge at Balpakram. Kipling's India is still alive if you look hard enough, and I intend to return there.

The Gul

There were these two cryptozoologists and a reporter in a Victorian explorer-themed pub talking about ape-men. The pub was Mr. Fogg's Tavern at St. Martin's Lane, Covent Garden, London. It was a tad expensive and, annoyingly, they had run out of cider, but the venue was a small conceit on my part. Between the three of us, we must have been at least as well travelled as Vern's adventurer.

Dr. Chris Clark has been on almost every expedition the Centre for Fortean Zoology has launched. For some years he had expressed an interest in visiting the Central Asian country of Tajikistan to search for relic hominins, possible offshoots of one of the ancestors of man or their relatives. In the 1950s, the Russian polymath Boris Porshnev had visited the country during the Soviet era with the Snowman Commission in the 1950s. He had spoken to many witnesses whose sightings had reached back into the nineteenth century. Chris had originally wanted to charter a helicopter to drop us at the top of one of the country's valleys, whereupon we would walk down on foot. This, however, had been too expensive.

Whilst researching online for the most recent sightings, I had stumbled across an article entitled "Tajikistan: Search for the Yeti" in an online magazine called *Standpoint*. The author was a journalist called Ben Judah, who had written the piece back in 2010. In it, he recounts visiting the Romit Valley and meeting with witnesses, people who claimed to have seen, and even been attacked by, hairy, man-like beasts whom they feared more than the mujahideen. His guide was chased by a female hominin ten years previously whilst he was searching for firewood in the mountains. He described a black-haired creature with dangling breasts. Travelling further, he talked to more witnesses. Another man attacked by a black-haired female creature and another who had a man-like beast attack his donkey. Another witness, a youth of fifteen, saw the creature, hairy and monkey-like, climbing over rocks just four days before. There are no monkeys in Tajikistan. The author could not decide if this was a living myth or a living creature. I decided to contact him.

And so we sat in the explorer-themed pub and talked hominins over gin and roast beef. Ben told Chris and me that almost every person he spoke to in the Romit Valley had seen one of the creatures or knew someone who had seen one. The local name for the beast was "gul," an Arabic word meaning to tear. In Middle Eastern folklore, these were desert-dwelling, man-like demons who emerged at night to feed on human corpses. It is from *gul* that we derive the word *ghoul*.

The creatures were generally described as man-sized, hairy, with monkey-like faces and a foul smell. They seemed much smaller than the yeti, more like the almasty of Russia. In turn, Chris and I told Ben of our own searches for the orang-pendek, the yeti, and the almasty, as well as other, non-primate cryptids. Ben was still undecided as to the nature of the gul, if it was flesh or fable, but he told us that, as he rose above the tree line and moved on to the Pamir Mountains, the stories of the gul vanished. I pointed out that if these stories were just make-believe, then they would have carried on to the communities who lived above the tree line in the barren wastes. A real creature needs food and shelter, however, and would of necessity be a forest dweller.

Ben initially wanted to come with us, but he sadly could not get the financial backing from any of the newspapers or magazines he wrote for, mindless celebrity chatter being apparently more important than scientific endeavour. However, Chris and I decided that our target area would be the Romit Valley, and we would be joined by Dave Archer, another stalwart of CFZ expeditions and an eyewitness to the orang-pendek of Sumatra.

With the most recent sightings emanating from the Romit Valley, it was there that we decided to head. In June of 2018, we found ourselves in the capital city of Dushanbe. We had been met at the airport by our guide, interpreter, and fixer, a young man named Daldat. After a night in a hotel, we made a quick visit to the museum in Dushanbe. There was a display of Neanderthal tools such as stone axes. There was also a reconstruction of a family of hominins. These were some of the strangest reconstructions I have ever seen. They had faces like Neanderthals but bodies like gorillas, down on all fours with bowed legs and ape-like feet. Leaving civilization behind, we travelled to the northeast and headed for the twin forks of the Romit.

The mountains in the area did not resemble the alpine-like peaks of my imagination. They were drier and brought to mind the mountains of Greece. They are well watered with rivers and streams, but the earth itself seemed dusty, stony, and dry. Nevertheless the area was highly productive—mulberries, plums, walnuts, cherries, apples, and pears all grew wild there. Bears, wolves, lynx, deer, and mountain goats all inhabited the area. Both forks had rivers running along them and small villages dotted their length. The mountains rose steeply on either side.

Eventually we reached our first camp area, close to a farm on the lower reaches of the lower fork of the Romit. Several tents had already been erected when we arrived. A young boy from the farm told us that his grandfather had seen a gul. It was covered in yellow and black hair. The boy could tell us no more as his grandfather was away. He would, however, be returning a few days later and we would be able to talk to him.

That night a spectacular storm broke. I had little sleep.

After breakfast the next day, we walked down the valley to the village of Tavish. On the way we met an old man walking the dusty path beside the Kafirnigan River. We stopped and, through Daldat, asked if he knew of the gul. The old man said he had never seen one, but he knew of them. We asked him to describe the creature to the best of his knowledge. His first words were

"Its thumbs are placed further back on the hand than a human's are." This may seem strange, but we were to hear this comment time and time again. The man went on to say that the gul was covered in hair, had long arms and a barrel chest, and was very muscular. He said that several people in the village claimed to have seen it.

At the village, Daldat asked around, and soon several men came forward to tell their stories. The first was a biology teacher called Raga Bali. About seven or eight years ago, he was camped some thirty kilometres up the valley. He and some others were cutting grass for livestock fodder. They were sleeping in tents. Raga Bali awoke to a noise outside the tent. It was light, and he thought it was morning. It transpired that it was moonlight, however, and the time was about three in the morning. Outside, the donkey was stamping and braying. Looking out, he saw a strange, hair-covered creature about twenty-three feet away. It was about five feet five inches tall and stood in a somewhat stooped position. Its eyes shone in the moonlight, and it had a monkey-like face. Its hair was black, and it had long arms. The thumbs were set well back on the hands and the fingernails were black. It looked muscular but not as massive as a gorilla.

The gul seemed to be trying to strangle the donkey with the rope used to tether it to a tree. The struggling ass broke free and the gul ran off. Raga Bali found man-like tracks on the ground where the thing had stood.

We showed Raga Bali a selection of pictures. These included a gorilla, an orang-utan, reconstructions of *Homo habilis*, *Homo erectus*, Neanderthal man, and *Australopithecus africanus*, and various illustrations of the yeti, sasquatch, and skunk ape. Instantly, Raga Bali chose Justin Osbourn's excellent cover illustration of Lyle Blackburn's great book on the Fouke Monster, *The Beast of Boggy Creek*. The picture shows a dark-haired skunk ape with yellow eyes slouching through a swamp. He was particular in saying that the hands were very like the hands of the creature he saw.

Raga Bali also told us that there was a stone shack, now abandoned, up the river. Years ago, an old man lived there alone. Raga Bali used to visit him in years past. During the night, something would throw rocks at the roof. The old man told him it was a gul.

A second man, Zai Dim, had an even stranger tale. He told us that he was delivering beehives. There are many honey farms along the Romit, and beekeeping is big business. In 1982, he was driving from a village further up the valley to take some hives to Tavish. It was around three o'clock in the morning. As he approached a wooded area, he saw a hairy animal that he took to be a bear run across the road in front of his car. It disappeared down a slope and into some trees. Zai Dim stopped the car and tried to get a closer look at the creature, but he could not see it. Thinking it had vanished into the trees, he turned to go back to his car. Suddenly, something grabbed him from behind.

Turning around, Zai Dim was faced with a creature covered in dark yellow hair. It slouched but stood on two legs, being about five feet five inches tall. It had a human-like face with wide cheekbones and slanting yellow eyes. The creature was female, with drooping, hairless breasts. The gul grappled with him and he saw that its thumbs were placed far back on the hands, and it gripped with its fingers alone. The creature wrestled him to the floor and pinned him down. He said it had foul breath. They struggled for about five minutes before he got an arm free and punched the creature in the face. It let go of him, and Zai Dim ran for his car and locked himself inside. The creature ran back into the forest. He said he was ill for weeks afterwards. Several other witnesses said this, and it could be a reference to post-traumatic stress.

Zai Dim felt that the attack had not been motivated by aggression, but because the creature had wanted to mate with him.

Again, when shown the illustrations and photographs, Zai Dim chose the Fouke Monster as closest to the creature he saw.

The third witness was a man called Gulmond. One morning, whilst it was still dark, he was walking along the Romit Valley in an area about thirty kilometres from Tavish. He was taking food to his parents, who were working on a farm. He noticed a figure walking behind him and assumed it was another person. Suddenly a hand grabbed his arm. He saw that it was not shaped like a human hand but had thumbs placed far back. Turning, he saw a female gul about five feet, five inches tall and covered in dark yellow hair of similar colour to a camel. The gul had drooping breasts and a foul smell.

The gul tugged at his arm and he pulled away. The creature kept on grasping at him and splashed him with water from the Kafirnigan. As the sun rose, it ran away. Gulmond, like Zai Dim, felt that the creature wanted to mate with him.

Like the other two witnesses, Gulmond selected the Fouke Monster illustration as being most like the creature he saw.

Sexual attacks on humans by apes are not unknown. Primatologist Birute Galdikas witnessed her cook, a Dyak woman, being raped by an orang-utan named Gundul in Borneo. She recorded it in her book *Reflections of Eden*.

> *I began to realize that Gundul did not intend to harm the cook, but had something else in mind. The cook stopped struggling. 'It's all right,' she murmured. She lay back in my arms, with Gundul on top of her. Gundul was very calm and deliberate. He raped the cook. As he moved rhythmically back and forth, his eyes rolled upward to the heavens.*

The famous bipedal chimp "Oliver," acquired by animal trainers Frank and Janet Berger, showed a sexual interest in Janet as he became an adult. Oliver even attempted to mate with Janet.

The strange shape of the hands was something I found very interesting. It was invariably the first thing the witnesses mentioned. If you were going to make up a story about a monster, the thumbs would hardly be the first thing you described.

The placement of the thumb is much more like that of a chimpanzee than a man. It is also reminiscent of the Australopithecines, a primitive subfamily of African-based hominids that flourished four to two million years ago.

More recent species, such as *Homo erectus* and *Homo habilis*, have a much more opposable grip, more like modern man. The feature may be a plesiomorphic trait—an ancestral feature retained by a modern organism. It could be a feature allowing the creatures to climb with more ease, much like the orang-utan with its reduced thumbs and elongated fingers.

When we returned to camp, we found the owner of the land, a man called She Rali, had returned. He was the grandfather of the boy whom we had met the day before. He was a park ranger and had several encounters with the gul over the past ten years. At first, he was the victim of rocks thrown at him when he was in his orchard. He never saw the assailant. Then one morning he saw an upright, ape-like creature looking up into a walnut tree. The creature ran away when it saw him. Another time he saw a female gul from only sixteen feet away. The creature seemed to point to his groin before running away.

Then, in June of 2017, he got an even closer look. She Rali was rerouting a stream to irrigate his crops when something grabbed him from behind and hugged him. When he turned around, the thing let go and fled into the forest. It was a female gul with dark yellow hair, long breasts, and a vile smell. It had a flat nose with a wide face and cheekbones. Again, the witness emphasized the thumbs being further back on the hands than human thumbs. He thought the creature was interested in mating with him.

When showed the cards he picked out a reconstruction of a yeti as being the closest to what he had seen, but he said that the hands were different.

We set up some camera traps in the area and baited the area with raisins, nuts, and eggs.

That night we had intended to stake out the area, but a fierce storm stopped us.

The witness Raga Bali offered to join us at the camp and assist in our hunt. We drove up the valley and found the now-abandoned shack he had told us about. Long empty, trees were now growing through it. It was a simple stone building, one story with a slate roof.

Raga Bali told us he had heard a tale about a woman who had lived in a remote house in the Romit Valley. Her husband had passed away and she lived alone. One night a male gul broke into her house and raped her, or so the story went. She later had a hybrid son, half-gul, half-human. The boy lived with his mother until her death. He was then taken in by relations in a town called Chuyangaron about twenty miles away. He was apparently a little slow, but otherwise normal. Apparently, he lived in the town still.

I was very sceptical of the story. I have heard human-hominin hybrid stories with the yeti in the Himalayas, the sasquatch in the US, the almasty in Russia, and the di-di in Guyana. However, I thought it might be worthwhile to try and find the youth.

The following day, Chris, Dave, and I were all taken ill and confined to the campsite. Though still unwell, we went to the village of Sorbu gi Dakana the day after that. Here we spoke with another witness, Mr. Aka Jon. The Tajiks are very hospitable people, and Aka Jon invited us into his house. We were given tea with bread and honey, as well as masses of cakes, sweets, and fruit. Oddly, the savouries, such as cooked goat and sheep meat, were brought out last. Stranger still, the Tajiks simply adore fried egg and chips. We were given it at almost every occasion that we were guests in someone's house. As we ate and drank, he told us of his experience back in 1978.

Aka Jon was out harvesting walnuts with friends in the Romit Valley. The group had made a fire and camped out. Sometime after they had retired, he had looked out of the tent flaps. He saw a male gul crouched by the fire, warming itself. It had long black hair and when it stood up, it was about as tall as a man. Its face was like a man's but broader. The neck was so short that it made it look as if the head sat directly on the shoulders. The creature smelled bad. When it saw him, it ran away.

He had heard tell that in the next village, back in 1956–57, there was a disabled man who had visited the forest on a regular basis to have sex with a female gul. One day he was found dead in the forest.

Back in the 1940s, a friend of Aka Jon had shot at a gul and missed. Some days later he was found dead in his home, and the locals believed that he had been killed by the gul in revenge.

That night, Dave and I went out into the hills behind the area where She Rali had his encounter. We used night-vision cameras in the hope of catching something on film. As we walked along, Dave said that he saw a creature's eyeshine in the trees to our right. Looking up, I caught a split-second glimpse of something large moving between the trees. It appeared to be a hunched figure covered in long grey hair. It vanished in an eye blink. Then, as we moved forward, the creature stepped out from the trees. It was a crested porcupine. What I had taken for long, grey hair in the brief instant I saw it was actually its long quills!

Back in the camp, Dave caught a large solifugid. Also known as wind scorpions or camel spiders, they are in fact neither scorpion nor spider, but a relation of both.

The following day we were mostly confined to camp with illness. We all had dysentery, and it was the very worst I have ever experienced in any part of the world.

In the morning we foolishly decided to visit some caves that were a number of miles along the river. We were intending to stay in them for a day and a night. The combination of heat and illness made me vomit violently. The mission was aborted, and we returned to camp. This was the sickest I've ever been on any expedition.

That evening, as we were feeling a little better, Raga Bali invited us for tea at his house. We met his brother and his partner, a French woman who asked us why we were in Tajikistan. When we explained what we were looking for, she became very frightened. She had not heard of such creatures before, and the possibility that they might lurk in the surrounding mountains seemed to genuinely disturb her.

We broke camp in the morning and travelled to Chuyangaron to see if we could locate the man who was supposed to be half-gul. All we had to go on was that he lived near a mosque. There were two mosques in Chuyangaron, an old one and a new one. We asked a young man close to the new mosque if he had heard of the story. He had not, but he said he could take us to the old mosque. He helped us by asking around close to the older mosque and found an old man who knew something of the story. He invited us in for lunch and told us what he knew.

The story we had been told wasn't 100 percent accurate. The man in question was not a youth; indeed, he was now dead. His name was Yattin, he had been born in 1956 and had died several years past, aged sixty. Yattin was supposedly half-gul, his mother having been raped by such a creature. He himself was totally normal. He married and had twins, who unfortunately died. Later, his wife had given birth to a daughter who was still alive. The daughter lived with a guardian in a suburb of the town, her mother having also passed away. Daldat got the details and we decided to visit Yattin's daughter.

Of course, three Englishmen couldn't just roll up, bang on the door, and say "Oi, was your dad half-human? Did his mum have it away with a relic hominin?" Daldat devised a plan whereby he would translate to the guardian that we were three of Yattin's old friends from England, come to pay our respects and meet his daughter.

And so we arrived at the house and were met by the lady who now looked after Yattin's daughter. She was very accommodating and introduced us to the girl. Her name was Moha and though she could tell us her name, she could not tell us how old she was. The guardian said she was nineteen. Moha looked fairly normal, thickset with a broad face and bushy eyebrows, but she clearly had nothing other than modern human genes. She did, however, have an intellectual disability.

The guardian brought out some old passport photos of Yattin himself. He too was thickset with a broad face, flat nose, and a thick, Brezhnev-type monobrow. He had a bushy black and white beard that made it look as if he had attached a badger to his chin. Yet, like his daughter, he was clearly a modern human. A hybrid with some kind of hominin would have shown primitive characteristics that neither Yattin nor Moha displayed.

The whole hybrid story could have been used to explain the intellectual disability of his daughter. In past centuries, deformed or disabled children in Europe were explained as "changelings." These were fairy babies that the little people swapped with human children. The human child would be taken away to Fairyland in order to bolster up the weakening bloodlines of the fay. The child left in its place would be an ugly, sickly child.

There is a real-life precedent, though. We now know that early modern humans did crossbreed with other hominins, including Neanderthals, Denisovans, and unidentified hominins only known from the genetic material they left in modern man.

The next day we broke camp and moved to the upper fork of the Romit Valley. We stopped at the first village, Qhyshan, and were offered a room in a lovely house on the banks of the Sardai-Miyona river. It was nice to have a roof over our heads.

In the village we met a local man called Nas Rullo. He had heard a story from the neighbouring village about a man who had been out fishing. On his way back home, he met a female gul who presented herself to him sexually.

Nas Rullo also mentioned that a tiger had been shot by a hunter in the valley. Only the year before, the man had shown him a picture of the tiger on his mobile phone. The authorities investigated but found no tiger.

The story, if true, was dynamite. Tigers did indeed once inhabit Tajikistan, but officially they had been extinct since nearly fifty years ago, the last one being killed in Turkey in 1970. The Caspian tiger (*Panthera tigris virgata*) was the second largest species of tiger after the Siberian. It had a distinctive long, thick coat and a ruff or short mane around the neck. The Caspian tiger lived in Central Asiatic Russia, Afghanistan, Iran, Iraq, Turkey, Mongolia, Azerbaijan, Turkmenistan, Uzbekistan, and Tajikistan. The idea that one was alive in the Romit Valley just one year ago was astounding. We decided to ask local people about the tiger as well as the gul.

We visited the mosque and spoke to a group of village elders, asking first about the gul and then the tiger. The men were very glad to help, and we gained much information from them.

We were told of a man named Zanadren who had an encounter around ten to fifteen years ago. He had been cutting firewood in the mountains. When he sat down, he was attacked by a male gul. It forced him to the ground, but he was able to hit it with an axe. The gul then fled.

Another story involved a shepherd who was tending his flock in the mountains. A gul appeared and blocked his way. The man struck the gul with a stick and killed it. If there is any truth to this story, I think this must have been a young specimen. A blow from a stick wielded by a human would not kill an adult chimpanzee. Apes are strong creatures with thick skulls and muscle mass. Most hominins from the fossil record seem to share these traits.

The body looked like a man's, but it was covered in black and yellow hair. He took other villagers to see the body. This happened in the Soviet era in a village now abandoned.

Another story happened around 1990. Two hunters went into the mountains but found no animals. They built a fire and made camp. Soon they were asleep. In the night, a female gul grabbed one of the men and clutched at his penis. He grabbed a burning stick from the fire and drove the creature off.

On another occasion, some men were drying yoghurt by suspending it in cloth hung from trees. They saw a large gul grab a bundle of the yoghurt and ran off with it.

One story featured a man who went into the mountains to search for a hunter who had vanished. He carried a gun for defence. One night, as he slept, a gul grabbed him and tried to drag him away by the legs. He managed to seize his rifle and shoot the creature dead.

The men did not know what had happened to either of the bodies in these stories.

They all believe that the gul is some form of wild man.

The elderly mullahs all said that tigers still existed in the mountains and hunted wild goats and Marco Polo sheep. One was said to have killed five domestic sheep in a pen about four to five years ago. It was seen by the farmer who trapped it in the pen. The tiger was killed by villagers. They did not know what became of the body.

About seven years ago, another man from the village saw a tiger. He described it as longer than a dog, with a tail three to four feet long. It was yellow with white and black stripes.

About fifteen years ago, a hunter saw a tiger kill a wild goat by biting it in the neck. The hunter scared the tiger away and took the goat, leaving only the head.

They insisted that these animals were not snow leopards. They knew that there were three big cats in the Romit, the leopard, the snow leopard, and the tiger.

We drove further up the Romit Valley. The trees grew sparser, and it became colder. On the road we met a man called Abdula and we stopped to speak. He claimed to have seen two guls. One he encountered about six years before whilst hunting with dogs. It was man-like and covered in black hair. The face was like a man's, but with a more protruding jaw line. The dogs attacked it and it defended itself by throwing rocks. It could run on all fours and upright like a man. It escaped by running away into the mountains.

His second sighting had occurred four years earlier. He was riding a donkey along the same road we were on when the animal stalled. He saw a creature hiding behind a rock. At first, he thought it was a bear, but then he saw it was a gul. It was covered in black hair and had a human-like face with a prognathous jaw. It loped off on all fours like a gorilla.

We made camp by the river. We were all still feeling rather ill.

Next day we hiked up the river valley. The scenery was beautiful, but the upper fork seemed more barren. We met up with a honey farmer called Asid. He invited us in for tea and honey. We asked him about the gul. He had not seen one himself, but his father had. His father had a machine for kneading dough. It was powered by the flow of the river. A female gul would sometimes come around and steal the dough. His father told him not to be afraid of her.

Tigers were a different matter; he had seen one himself. He had been hunting with friends and had shot a wild goat. A tiger appeared and took the dead goat away. Last year a hunter had told Asid he had seen tigers twice. In both cases, they were females with cubs. One was a group of six, another a group of eight.

We walked on till the end of the gorge. We retraced our steps and then clambered up some steep rocks. Above this was another ridge, and above this, another. In the burning heat and still suffering dysentery, this was too much for me, and I had to turn back, exhausted.

We visited the village of Vishtan the next day. We spoke with an elderly mullah who still went hunting despite being seventy-four. He had never seen a gul, though he had heard stories of them. He had seen tigers on three occasions, all of which were before the civil war (1992–1997). The first two times, he had seen tigers crouched in the undergrowth, as if in ambush mode. The third time he saw one hunt and kill a deer.

In the next village, we spoke to another old mullah who was also a hunter and a beekeeper. His name was Bobo Safa. His father and grandfather warned him about guls in the mountains, but he never saw one. Twenty years before, he had seen a tiger in the Romit from a distance of no more than sixty-five feet. He had heard of sightings of females with cubs. He had also heard a story of a tiger that had been killing sheep and had been trapped in the sheep pen by villagers.

Later that day, we spoke with a park ranger called Namon. He did not want to be filmed or photographed but he told us of what he had seen. At around ten in the morning on June 18, 2018, he had seen a Caspian tiger.

He was high in the mountains and there was still snow on the ground. He estimated that the tiger was a young adult, about three or four years old. When the animal saw him, it left. It was the only time he had ever seen a tiger in the wild.

The following day, our illness escalated and confined us to camp. Dave caught two venomous red-backed spiders close to the camp. These beautiful black spiders with red markings are related to the more familiar black widow spiders of North America.

We were well enough to go along the valley the next day. We took a look along a stream and traced it back to a waterfall. We checked for any bones in the water but found none. On the way back, we took tea with a group of honey farmers. One of them, Achmed, had seen a tiger back in the Soviet era (1929–1991). It had been in the Pamir Mountains to the east of Romit. His father had once caught one and sent it to a zoo in Dushanbe. More recently he had heard of a tiger killing twenty sheep.

A young beekeeper called Kaseem had seen a pair of guls a few years ago. He was working with another man in a water-powered mill in one of the streams upriver that led down to the Sardai-Miyona. The mill wheel stopped turning, so Kaseem went upstream to see what the blockage was. He discovered two man-like creatures sitting in the stream and blocking the flow. They appeared to be a male and a female. The female was human-sized, the male somewhat larger. They had human-like faces and were covered with black hair. They had a foul smell. As soon as the creatures saw Kaseem, they became aggressive. They chased him back to the mill. He and the other man locked themselves inside the mill as the creatures banged on the door and leapt up onto the roof. The creatures prowled around the mill for an hour. The second man, a mullah, tried to calm Kaseem down as he was panicking. The other man claimed to have seen the creatures before.

Back at camp, Raga Bali told us that he too had seen tigers, about seven or eight years ago near the village of Tavish. On the first occasion, he had seen a female with three cubs on the far bank of the river. They were all feeding on a dead deer. He watched them feed for an hour. The second time, he saw a single tiger wandering along on the far bank of the river. He thought that they came down from higher elevations in winter.

Before breaking camp, we retrieved the camera traps.

And so the expedition wound down. I was ill for three weeks after I returned to England. Checking the trail camera pictures, we found that we had captured bears, foxes, and wild boar, but no hominins.

I wrote to as many tiger conservation groups and organizations as I could with the information of sightings of the Caspian tiger in Tajikistan. The silence was deafening. Out of all of them, only one group bothered to grace me with an answer, and they said that they were only concerned with the Siberian tiger. I was both surprised and disappointed that none of these groups found these accounts interesting enough to bother with. The idea of such a spectacular predator, long thought to be extinct, actually surviving in modern-day Tajikistan is both exciting and fascinating. I have tried to disseminate the information elsewhere.

So what are we to make of the gul? Before I visited Tajikistan, I thought the creature would be the same species as the almasty of Russia. However, the two seem different. The almasty could, according to witnesses, reach seven and a half feet tall. The gul was more like a man in height, if far broader across the shoulders. More telling is the strange structure of the hand. All the witnesses stressed that the thumbs were further back on the hand than a human thumb. If you were going to make up a story about seeing a monster, would the thumbs be the first thing you described?

The hands of fossil hominins such as *Homo erectus* or *Homo habilis* seem more like modern man in structure. Even the more primitive australopithecines had a hand structure more man-like than ape-like. The shape of the hands of the gul as described by witnesses looked more like those of a chimpanzee or those of *Ardipithecus ramidus*, a 4.4-million-year-old hominin that was twice as ancient as *Homo habilis*. Does this mean that the gul is a descendant of *Ardipithecus ramidus or* one of its relations? Possibly, but not necessarily—the strange hand shape may be a relatively recent development, perhaps an adaptation to climbing, but all this is just speculation. Only a specimen will answer these riddles once and for all. It seems that the gul may be a whole new chapter in hominology.

CHAPTER FOUR

How to Hunt for Monsters: Organizing a Cryptozoological Expedition

"The real world is where the monsters are."

—Rick Riordan

You might thing that trekking through steaming jungles and swamps or crossing deserts and mountains on the track of monsters was something confined to the pages of pulp novels and *Boy's Own Adventures* of the 1920s and '30s. But you would be wrong!

A small but dedicated group of people around the world, cryptozoologists, brave the dangers of the unexplored corners of our planet on the track of unknown animals. I am one such person. As the zoological director of the Centre for Fortean Zoology, I have travelled the globe trying to track down these beasts of legend. This is not a hobby for me, but a job.

In this chapter, I will explain how to mount a cryptozoological expedition of your own. I hope to inspire others to take up the challenge and hit the trail. If successful, scientific immortality awaits you.

If you are thinking of going monster hunting for profit, think again. No one got rich through this game. Cryptozoological expeditions devour money. Cryptozoology is a calling; it's about dedication, adventure, and scientific endeavour, not profit.

Organizing a monster hunt might sound daunting, but if you break it down into steps, it is not really so hard. To be honest, the most difficult things I have dealt with on expeditions are getting to and from the airports across London.

The main thing to remember is the six *P*s: perfect planning prevents pretty poor performance!

Choose Your Monster

First of all, you will need to select a cryptid that you want to search for. You don't want to waste time chasing shadows. What makes a good target creature? There are several factors to consider. Is your target creature likely to exist? It makes sense to look for a creature that has been reported by witnesses for a very long time rather than something seen only once or twice. In some cases, such as the Canadian sea serpent dubbed cadborosaurus, the Sumatran mystery ape orang-pendek, and the thylacine or marsupial wolf of Tasmania, some of the witnesses have actually been scientists. Such witnesses strengthen the case for the existence of the animal in question.

Another consideration in choosing your monster is the location. This is tied up with other factors such as finance and transport. For example, in order to take a decent expedition into the Congo, on the track of Mokele-mbembe, the supposed living dinosaur (in fact far more likely to be a giant monitor lizard) would be an expensive undertaking. In order to get there and transport yourselves, bearers, and equipment into the very deep jungle you would need a large team of people. You would need to weigh this against the noise such a large group would make and its possible negative effects.

In an expedition to the wilds of Siberia after the Irkuiem or god bear, you would be hundreds of miles from the nearest town. You would need to take enough provisions to last for the whole time you were there. You would need to think about how you would transport this or if you were going to store it at a base camp.

Try to select a creature you have a fighting chance of seeing or at least gaining some information on from eyewitnesses.

Don't forget to bone up by reading as much as you can on both the creature and the country it inhabits. Guides such as *Lonely Planet* are excellent to give you travellers' info on almost any country.

Planning Ahead

Always have an itinerary for your expedition. You will only have limited time and you want to make the most of it. Try to get information on where the creature was last seen, and concentrate your searches on that area. However, be ready to move on if the monster turns up elsewhere whilst you are in the country.

Oftentimes guides will do some groundwork in advance for you, such as finding witnesses for you to interview. With e-mongol.com, the CFZ designed posters that were then translated into Mongolian and distributed to nomads in the desert. They explained how a group of British scientists were visiting the area and wanted information on the death worm, or even a specimen. Though no specimens were forthcoming, the posters attracted many witnesses, some of whom travelled several days to see us.

In Russia, our Ukrainian and Russian counterparts had been in the area for two weeks chasing up stories and finding eyewitnesses.

You might want to set up camera traps in the hope of photographing your quarry. Test your trigger cameras before setting them up in the wild. The same goes for all your equipment.

Funding

Sadly, there are no academic bodies that finance cryptozoological expeditions. The Centre for Fortean Zoology generally finances its own expeditions. We publish our own books, do lectures, and write for magazines, as well as publishing our own quarterly magazine, *Animals and Men*. The money from this goes toward expeditions. Members generally pay their own way on such trips.

If you have a day job, you might want to save up and take your annual fortnight off on a monster hunt. Remember, though, this is not a holiday. Cryptozoological expeditions can be dangerous and often take you to remote places. A monster hunter needs to be brave enough to follow the quarry into places most people would never dream of going.

Expenses can vary greatly. This can include flights, equipment, and native guides. The cost of my trip to West Africa on the track of the dragon-like ninki-nanka was a little over £300. My Russian trip looking for relic hominids known as almasty ran to about £1500. More expensive expeditions have included my hunt for the Mongolian death worm in the Gobi and the giant anaconda in Guyana, costing £4,000 and £7,000 respectively.

You might think TV companies would be interested in financing cryptozoological expeditions in order to make exciting documentaries out of them. Nothing could be further from the truth! Of all my expeditions, only one, the 2000 trip to Indo-China in search of the serpent dragon called the naga, was financed by a TV company. About six or seven time per year, the CFZ are approached by researchers for companies that are toying with the idea of making a film about an expedition. They all end up the same way. The researchers go back to the company, the company approaches the TV station with the idea, and the TV station says no. Usually there is no explanation, but in the past, we have been told we were rejected for being too real (too much like natural history and not fitting in with dross like "alien abductions" and "rescue mediums"). On another occasion, we were contacted by a researcher who wanted to send a camera crew along on a CFZ-funded expedition in order to make a documentary for the BBC. In return, he said, we would get nothing! What I said to him is quite unprintable here. TV companies promise the earth and deliver nothing.

We were once funded by a video game company, but this was the one-off whim of one man who worked for the company in question. On the whole, cryptozoological expeditions are funded from the cryptozoologists' ever more threadbare pockets.

But wait, I hear you cry, *I can't afford to fly to the other side of the world!* Don't worry—sometimes monsters are on your doorstep. An investigation in the UK can be done for next to nothing. The CFZ have hunted big cats and lake monsters all over the UK. In fact, our most successful expedition was to Martin Mere in Lancashire, where we investigated reports of a monster attacking full-grown swans. It turned out to be a giant catfish.

You don't have to clamber up the Himalayas or delve into the Amazon to be a cryptozoologist. There is a wealth of poorly investigated creatures in the UK and Ireland, such as the earth hounds in Scotland and the master otter in Ireland.

Your Team

If you intend to take a group of people on your expedition, you need to select them with care. Don't bring just anyone along. Ask yourself if the people you have in mind will be physically and mentally up to the job. I recall on one expedition to Africa, one man, who claimed to have much experience in the Dark Continent, was of no use whatsoever once he was there. This "old Africa hand" spent most of the trip in a seedy hotel room whining about the humidity whilst the rest of us traversed swamp and jungle.

Look for people with zoological experience or tracking experience. Those well versed in the use of cameras will also be a boon to your team. Most of all, make sure all members can get along together well. You don't want fights breaking out whilst you're up a mountain or in a rain forest.

Academic qualifications are not always a guarantee of quality. Whilst studying zoology at Leeds University in the 1990s, I was appalled at the lack of knowledge and infantile errors from people who were supposed to be teaching the subject. A degree is no match for practical experience and fieldwork.

Don't make your team too big, unless you plan to split up. Most of my expeditions have consisted of three to five members. Remember, you will have guides as well. If you get too many people tramping through the wilds like a herd of demented elephants, any self-respecting cryptid will hear you coming and vacate the area post haste!

Native Guides

This is perhaps the most important factor in any monster hunt. It is vital to get hold of trustworthy and experienced local people to help you in your expedition. I have been lucky in having worked with some of the best guides in the world. Some companies will provide excellent guides with intimate local knowledge. When we investigated the Mongolian death worm, we worked with the brilliant Mongolian company e-mongol.com.

They provided us with English-speaking Mongolian guides and drivers who knew the desert and the nomads like the backs of their hands.

In Guyana, we were lucky enough to have a native chief of the Eagle Clan Arawak Indians as our guide. Damon Corrie of guidedculturaltours.com took us off the beaten track and into native villages. We were honoured to be the first Westerners allowed into caves where an ancient burial had taken place. Thanks to Damon, we uncovered information on monsters unheard of outside the country. Getting away from tourist areas is one of the keys to carrying out a successful expedition.

Sometimes your guide will be as interesting as the monster you are hunting. In Sumatra, our guide Sahar was a shaman said to be able to bring down the "tiger spirit" to possess him. In the jungle, his bushcraft was second to none, and we even found and followed the tracks of an orang-pendek. His friend Debbie Martyr, head of the Indonesian Tiger Conservation group, had seen the orang-pendek several times.

Often the guides will not only know monster witnesses, but will have seen the creatures themselves. An elderly man in Thailand took me into a maze of caverns in the jungle to show me where he saw a sixty-foot naga crawling through a subterranean river. In Russia, the respected Ukrainian scientist Gregory Panchenko had twice seen the almasty, once at very close range.

Transport

Getting around in developing countries can be an adventure. In some countries, you will have to transverse vast distances. In my experience, there are two types of expedition. The first is when most of the exploration is done on foot due to the terrain. You may drive to the area initially, but then you spend weeks in the jungle or mountains. The second is where you drive from place to place and spend comparatively little time walking. Sumatra, Russia, and Guyana fell into the former type, and Thailand and Mongolia into the latter.

Roads, where they exist at all, tend to be much worse than in the UK—so full of potholes that driving along them can be as dangerous as hacking your way through the rain forest. Mountainous areas like the Caucasus and

Sumatra have tortuously twisting roads, making a journey that would last two hours in the West last up to eight hours.

I recommend you get a four-wheel-drive vehicle of decent size. This way, you can use it to carry equipment and even to sleep in if need be.

Inoculations, Tablets, and Health

Before you go tearing off on a monster hunt, make sure you have the correct inoculations, or you might be falling victim to much smaller monsters! Make an appointment at your local health centre. They will have detailed information on the inoculations you need for each country. Some areas will differ from others. Visitors to Mongolia will need a series of injections against rabies. The neo-tropics, South and Central America, harbour yellow swamp fever. Most tropical areas will have malaria-carrying mosquitoes. Malaria tablets are a must in these cases. Depending on the kind of malaria tablets you are on, you may have to start taking them several weeks before you leave.

Always bring Imodium for an upset stomach and water purification tablets. The latter are available at most camping/outdoor stores. This will kill off most bacteria in drinking water, enabling you to refill your flask from lakes and streams. Always read the instructions carefully. Decent insect repellent is a good idea, as is a powerful sunscreen for hot countries. A first aid kit with plasters, antiseptic cream, and bandages is a must. If your first aid kit contains scissors, remember not to carry them in your hand luggage on the flight!

Obviously, make sure you have comprehensive travel insurance before you leave.

Equipment

What you need to take can differ radically from expedition to expedition. Generally, you will be camping out in the wilds. Good, hardy footwear is essential. I once knew a man who came on an expedition with tatty old boots whose soles fell off three days into the trip! He spent the rest of the arduous expedition in soft slip-ons!

Make sure you know about the climate of the country you are visiting. Remember, many countries have rainy and dry seasons. If you are going to a wet climate, you will need a rain poncho and a waterproof bag to keep your extra clothes dry.

Take warm clothes for a cool climate and lighter one for a tropical country. Remember that even in the tropics, it can get cold at night. I recall on my first trip to the mountains of Sumatra I was freezing cold at night because I had not brought an adequate sleeping bag. A good multi-season sleeping bag is a must.

Sometimes guides will provide tents and camping equipment, but don't bank on this—enquire first. Guides usually double up as cooks and have their own cooking equipment. Sometimes the guide's fee covers food expenses, sometimes not—make sure this is all worked out in advance.

You need to travel as light as you can. If your expedition is to be made mostly in vehicles, than this is not too much of an issue, but on foot, believe you me, you can feel every ounce in your backpack. Apart from tents and bedding, take as few clothes as you can. You will end up smelling like a tramp, but this is a monster hunt, not a fashion show! You will need to leave room for other equipment and any evidence you find as well. If you carry too much, there will be an additional charge on flights.

Always take both moving and still cameras. Imagine if you saw a cryptid and had no camera on you! Who would ever believe you? I find digital cameras to be best. Take extra batteries, as electric sockets for recharging are few and far between in jungles, deserts, and mountains. Trigger cameras or camera traps can be picked up relatively cheaply online. Remember where you set the traps up. You don't want to forget where you put the camera and lose it in a forest! In the UK, there is always the risk of the camera trap being nicked, so chain them in place. The CFZ has had one camera trap swiped when it was put up in a field to photograph a big cat.

On my last trip to Sumatra, I found that the excessively damp cloud forest drained the batteries of my camera very quickly. Try to store extra batteries in waterproof bags. Take as many as you can.

Also carry scientific specimen bags for any skin, hair, or dung you may find. These bags are easily bought online. When handling evidence, always use surgical gloves so as not to contaminate the DNA with your own. A good resin for making casts of footprints is a nice idea. Resin is generally stronger and keeps the shape better than plaster of Paris. I remember carrying a plastic bag full of plaster of Paris through customs and worrying what they would make of my bag of white powder! In very damp climates, silicone rubber (available from DIY shops) is the best bet for creating a good, waterproof, and tough cast.

Sound recording equipment, such as a powerful Dictaphone, is good for capturing vocalizations of creatures and eyewitness statements.

How to Make the Best of Your Time

If you are taking a big group, then split up to cover more ground. Smaller groups also make less noise. Record all interviews with witnesses and film as much of the expedition as you can. This will come in use not only as a record in itself, but also as a tool to improve future trips.

You will be very lucky to get a glimpse of your target creature, so gleaning information from locals and witnesses is the next best thing.

It is a good idea to prepare a list of questions when interviewing a witness. Be patient, as English will probably not be their first language. You will have to work through an interpreter. In Sumatra, when speaking to the Kubu tribesmen of their encounters with orang-pendek and giant snakes, we needed two interpreters. Our guide Sahar translated our questions into Indonesian, and then these were translated by a second man (who spoke no English) into the language of the Kubu and vice versa.

Generally, I have found native peoples to be very hospitable and honest. They are usually surprised at scientists from "the outside world" being interested in their sightings. They ask for no payment, and often do not consider what they have seen to be anything special but just "another kind of animal." One Russian almasty witness was amazed that we were interested in such a "crazy topic," as she put it.

The only real exception to this was in the Gambia, where everyone seemed to be out to twist money from Europeans and would tell any kind of tall tale if they thought they could get their grubby hands on money.

Always keep a journal of your trip, with entries every day. Such notes are invaluable when it comes to writing an account of your expedition. In the modern day, you may even be able to do a daily computer blog with a laptop if you can get a signal.

What to Do When You Get Home

If you have gotten lucky and managed to film a cryptid, do not hoard the film and exchange it only for money. As noted before, this is a scientific endeavour, not a way of making cash. Share the film with fellow cryptozoologists and the scientific community. This is all about exchanging knowledge and advancing knowledge. If you hold out for cash, people will be suspicious. If you have indeed filmed an unknown animal, then your reward will be scientific fame.

If you have bone, skin, dung, scales, or other organic matter that may come from a cryptid, you will need to get it scientifically analyzed. Approach reputable museums, universities, or zoological organizations. If these cannot do the analysis themselves, they may be able to furnish you with contacts who can. Most of the organic matter from CFZ expeditions is analyzed by Copenhagen University, where our friend Dr. Lars Thomas works.

Whatever the outcome, publish the findings. There is no shame in mistaking something mundane for something strange. Supposed orang-pendek hair the CFZ brought back from Sumatra on our first expedition there turned out to be from a golden cat. Possible almasty hair was in fact human. Such negatives do not mean your monster does not exist, just that you don't have a part of it. It's better to have made a genuine error whilst looking for a cryptid than to never have looked for it at all. One day, someone will get lucky and come up with the goods. That someone might just be you!

On our latest trip to Sumatra, we brought back possible orang-pendek hair. The DNA has been analysed by Lars and his team and he found it was similar to, but not the same as, orang-utan DNA. At the time of writing, a further set of tests is being carried out. It looks as if the hair may belong to a new species of ape.

Always record your findings and your trip in a write-up. This can be as a book, a magazine article, an online account, or, as we at the CFZ do, all three. Recording your work for posterity is vital. You would not want the things your expedition discovered to be lost and forgotten, would you? By recording them, they will be here forever and be a useful resource to other cryptozoologists.

The Best Way to Hunt Monsters

Finally, the best way to have a chance to hunt real-life monsters in the twenty-first century is to join the only full-time, professional cryptozoological organization in the world, the Centre for Fortean Zoology. The CFZ organizes monster hunts all over the world. Check out our website, www.cfz.org.uk.

Even now, monsters still stalk the earth, and great discoveries are waiting to be made. Adventure is just around the corner, if you're brave enough. Now it's all up to you.

An Afterword

And so we have reached the end of our journey. To be honest, this book has hardly scratched the surface of this subject. I could have written a series of books, each of encyclopaedic length, on cryptozoology. I hope, however, that I have piqued the reader's interest.

The evidence for many of the creatures covered in this book is now mounting, but mainstream science still remains largely hostile to the subject. My friend Adam Davies once prepared a paper with Professor Hans Brunner, a world-recognized expert in mammal hair. The subject was their findings after the analysis of suspected orang-pendek hairs from Sumatra. The paper was submitted to the magazine *Nature*. The editor refused to publish the paper on the grounds that it was dealing with a large unknown animal. He told them that, if it were about a new species of mouse, then there would have been no problem! What an unscientific and ridiculous approach to data. The attitude beggars belief.

It seems that, if a theoretical physicist writes a paper on a molecule that nobody has ever seen, but its existence is inferred by the reactions of other molecules, then that is fine. But should a zoologist write a paper on a creature that has been seen by thousands of people, but for which we do not yet have a type specimen, then he is a scientific heretic. These double standards can do nothing but harm science in the long run.

Remember the fate of the sorely wronged Pierre Denys de Montfort? Little has changed two hundred years since his death. It is up to people like you and me, dear reader, to change that. As Bernard Heuvelmans once said, "The great days of zoology are not done." The discoveries are there to be made, in the jungles, mountains, deserts, and oceans, not in the lecture halls. There are still wild and unexplored places on Earth, and there are still unknown creatures inhabiting them. When your mother and father told you there are no such things as monsters, they were wrong.

—RICHARD FREEMAN,
EXETER, UK, JANUARY 2019

About the Author

Richard Freeman is a former zookeeper who has worked with over four hundred species of animals and has a special interest in crocodiles. He is a full-time cryptozoologist and is the Zoological Director of the Centre for Fortean Zoology, the world's only professional organization dedicated to searching for unknown species. He has searched for cryptids on five continents and has investigated creatures such as the yeti, the Tasmanian wolf, the orang-pendek, the giant anaconda, the Mongolian death worm, the almasty, the ninki-nanka, the gul, and many others. He is currently planning a series of trips in search of giant, man-eating crocodiles. He has lectured widely on cryptozoology at venues such as the Natural History Museum and the Grant Museum of Zoology. He has written a number of books on cryptozoology and folklore as well as horror fiction. His interest in strange creatures stems from a love of the BBC science fiction television classic *Doctor Who*.